T0355605

Bodies and Mobile Media

Bodies and Mobile Media

Ingrid Richardson
Rowan Wilken

polity

Copyright © Ingrid Richardson and Rowan Wilken 2024

The right of Ingrid Richardson and Rowan Wilken to be identified as Authors of this Work has been asserted in accordance with the UK Copyright, Designs and Patents Act 1988.

First published in 2024 by Polity Press

Polity Press
65 Bridge Street
Cambridge CB2 1UR, UK

Polity Press
111 River Street
Hoboken, NJ 07030, USA

All rights reserved. Except for the quotation of short passages for the purpose of criticism and review, no part of this publication may be reproduced, stored in a retrieval system or transmitted, in any form or by any means, electronic, mechanical, photocopying, recording or otherwise, without the prior permission of the publisher.

ISBN-13: 978-1-5095-4961-0
ISBN-13: 978-1-5095-4962-7(pb)

A catalogue record for this book is available from the British Library.

Library of Congress Control Number: 2023934598

Typeset in 11 on 13pt Sabon
by Fakenham Prepress Solutions, Fakenham, Norfolk NR21 8NL
Printed and bound in the UK by TJ International Limited

The publisher has used its best endeavours to ensure that the URLs for external websites referred to in this book are correct and active at the time of going to press. However, the publisher has no responsibility for the websites and can make no guarantee that a site will remain live or that the content is or will remain appropriate.

Every effort has been made to trace all copyright holders, but if any have been overlooked the publisher will be pleased to include any necessary credits in any subsequent reprint or edition.

For further information on Polity, visit our website:
politybooks.com

Contents

Acknowledgments

We wish to thank Mary Savigar and Stephanie Homer at Polity for their patience, support and encouragement. Ingrid dedicates this book to her home team: Zoe, Jamie, Tig and Izzie. Rowan dedicates this book to Karen, Lazarus, Max and Sunday.

All watermark images at the beginning of each chapter are sourced from Wikimedia Commons.

Introduction

Come now, with all your powers discern how each thing manifests itself, trusting no more to sight than to hearing, and no more to the echoing ear than to the tongue's taste; rejecting none of the body's parts that might be a means to knowledge, but attending to each particular manifestation. (Empedocles, cited in Ihde 2007: 8)

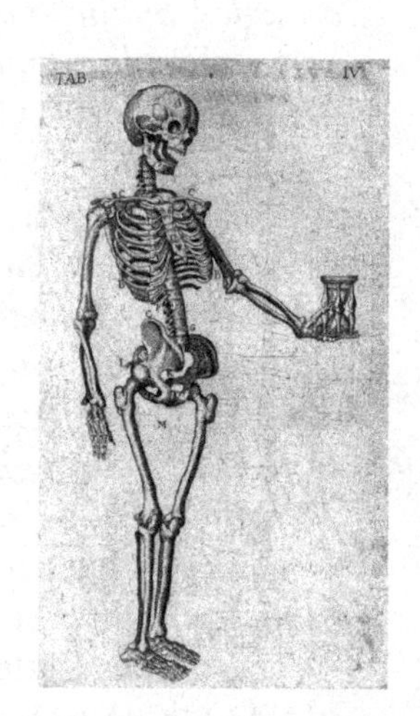

Zara has been playing the augmented reality location-based game Pokémon GO for nearly seven years, since the mobile app was first released in 2016 and she was in high school. She's a bit old for it now, she thinks, but it has become a comfortingly familiar and almost incidental part of her everyday life, infused with the nostalgia of the Pokémon universe she explored as a

child playing boardgames and the Nintendo DS with her friends. The game is an all-of-body experience, requiring deft gestural flicks and swipes with her fingers on the haptic mobile screen, and visual oscillation between the screen and physical environment as she encounters and captures the pocket monsters. At times she likes to hear the ambient sound effects or re-experience Junichi Masuda's signature soundtracks – though the audio quality is not that great, so she often deliberately mutes the sound, especially in crowded public spaces where she needs to engage in some complex "facework" to avoid bumping into the other people and objects surrounding her. In the early days of the game, she would take the family dog on long walks with the dual purpose of hatching Pokémon eggs, but now that she's working, she'll occasionally "cheat" by exchanging codes with her friends or customizing speed with a location changer app. For many years she has modified her walking trajectories through the city and her neighborhood, ensuring she encounters the most PokéStops to collect valuable game resources, and Gyms to engage in multiplayer battles and win game currency.

There is an intimacy in these hybrid and playful place-making practices, and a very personal sensory connection to a device that is both perpetually on-hand and online, affording an augmented real-time experience of the world that comingles the sights and sounds of place and interface, and entangles her hand–eye coordination and pedestrian footwork with the mobile touchscreen.

With every change to our technological interfaces, there is a corresponding modification to perceptual reach and communicative possibility. The shift from analogue to digital technologies and forms of media over the past fifty years has mobilized a critical transition in how we relate to and make meaning of the world. One of the most significant cultural effects of this translation of image and information into digital code is the increasing predominance of mobile and haptic screens.

The mobile device is simultaneously – and often equally – an aural, visual, haptic and peripatetic interface, requiring an evolving sensibility and a cross-modal literacy of which our bodies, in their perceptual flexibility, are inherently and essentially capable. Mobile media usage is quite literally a way of "having a body" that demands a complex adaptation. Indeed, our use of mobile media can be described in Leder's terms as an ongoing incorporation by which we reshape the "ability structure" of our bodies (Leder 1990). In these terms, then, how does this portability, along with the prosthetic and orthotic capacities of mobile media, impact upon our experience? How can we critically interpret the particular modes of embodiment emerging from our mobile media practices? What is the "status" of each of the senses with regard to these practices, and how have they changed our bodily engagement in and with the world?

The opening vignette describes the way mobile media – as navigational, playful, communicative, entertainment and creative devices – are thoroughly "embodied" and integral to our perception and experience of the world. Mobile media comprise all media carried or worn on the body, each effectively "mediating" our experience of the world; early forms include newspapers and books, though our focus in this book is primarily on digital media. Many of our examples involve the smartphone, as this is the most ubiquitous and familiar or mundane form of mobile media, but our descriptions also extend to eyewear, watches and tablets.

In this book, we explore this body–technology relation by tracing the emergence of what has been termed the "corporeal turn" in critical cultural theory. Our key aim is to *translate* this corporeal turn, by providing interpretations of mobile media theory and narratives of mobile media use in terms of specific sensory ratios or body–technology couplings – that is, where particular senses or parts of the body are privileged over others depending on the use-context. In this way, the book is intended as a form of knowledge translation in providing detailed accounts of how we interact with various media forms, especially mobile media. This knowledge translation, we argue, is important for us to critically understand how our perception and experience of the world, while partially configured by sociocultural context, is at the same time *mediated* by technology.

The corporeal turn is grounded in a phenomenological perspective, a branch of philosophy that directly challenged Descartes' famous adage "I think therefore I am," which positioned the mind and the intellect as that which grants us our awareness and defines the essence of what it means to be human. In phenomenology – particularly through the work of Maurice Merleau-Ponty – knowing and being are primordially grounded in sensory perception, in the fact that we have the kind of bodies that we do: bodies that see, hear, smell, taste and touch in synesthetic ways, and move about in a world of gravity and matter. Sensory perception is also always contextual and situated, intertwined with the objects around us, including not least our own tools and technologies. Merleau-Ponty and subsequent phenomenologists such as Don Ihde therefore argue that the body–technology relation is "primary" in terms of configuring our experience of the world – a position that we also take in this book. As each body–technology coupling has its own unique affordances, here and in the chapters that follow we argue that it is worth understanding in detail the relationality specific to mobile media. As an "intimate" technology, the mobile interface has had significant perceptual and experiential effects on our knowing and being in the world. Our approach to this endeavor is to adopt a phenomenological method, which is a mode of *discovery-through-description* of everyday human behavior. While a number of scholars have applied a phenomenological or "embodied" approach to mobile media practices, research in the field is fairly recent and spans diverse disciplines, and much of this work is dispersed and not focused in granular ways on specific senses. In part this book adopts a deliberate strategy of "gathering" existing work on sense perception and the body–technology relation as it pertains to the mobile interface, including relevant histories that have privileged some sensory modes above others.

We focus specifically on mobile media as an effective lens through which to interpret the *multi-sensoriality* of media use, incorporating all the senses, including the body–technology couplings of face–interface, eyes–screen, ears–sound, hand–touchscreen and pedestrian–device mobility. Our conceptual approach, while acknowledging this multi-sensorial engagement with mobile media, considers the "privileging" or uneven "ratio" of sense perception in mobile media practices, with individual chapters focusing on the face, eyes, ears, hands and feet. This focus on the specific senses allows us to both articulate and critique how there is a sensory hierarchy at work in our use of media – how certain senses have more status than others in various media practices. This hierarchy is also historically sedimented in our experience of media interfaces more broadly, such as the privileging of vision in the ocularcentric evolution of screens. There are different dimensions or modalities of use that can be interpreted in terms of the specific sensory affordances at work.

In this book, we examine the close relation between bodies, sensory perception and media use over the past few decades, up to and including contemporary mobile media practices. In the passage cited at the outset of this chapter, Empedocles encourages us to pay attention to each "particular manifestation" of the body's sensory apparatus – that is, to specific body parts and bodily senses. Focusing on each modality of sensory perception that is particularly significant to our use of mobile media allows us to drill down into the specific biases and effects in a way that would not be possible if we were to begin with a multi-sensory or synesthetic approach; but, importantly, it is also a "heuristic" or interpretive strategy for producing fine-grained critical descriptions of the many collective and idiosyncratic ways in which we engage with mobile devices, across different abilities, habits and literacies.

Each chapter thus has its own sensory foci, though the discussion will also address the relationality with other senses and sensory affordances where relevant. Structurally, the chapters begin with a detailed biological description – of the face, eyes, ears, hands and feet – an intentional "mooring" tactic that situates or saturates the reader in anatomical minutiae, as a phenomenological reminder of the body as "ground." This biological description then extends out to the more socio-somatic or cultural aspects of each mode of perception, before focusing on the body–technology coupling specific to mobile media, including helpful vignettes of scenarios of use that reveal the variability of both our bodies and our media practices. The concluding chapter brings the chapters together in a synesthetic approach, in addition to touching on modes of sense perception that are more peripheral to or emergent within mobile media use (such as taste and smell). In the following sections of this introduction, before giving a brief overview of the chapters, we offer some explanatory framing of the corporeal turn, phenomenology and postphenomenology, and the significance of metaphor in our embodied experience.

The Corporeal or Sensory Turn

In the latter part of the twentieth century, within cultural theories such as poststructuralism and postmodernism, there was a dedicated focus on language and "the text" as our primary modes of knowledge-making and communication. In reaction to this "linguistic turn," a subsequent materialist and corporeal turn (Sheets-Johnstone 2009; Tambornino 2002) sought to recognize the reciprocal relation between texts and "things," including bodies, and all the material, physical and sensory aspects of knowledge and being in the world. Within the corporeal turn, Ruthrof (1997: 7) argued, "meaning has ultimately to do with the body and … linguistic expressions mean anything or nothing at all unless they are activated by haptic, visual, tactile, gustatory, olfactory, and other non-verbal signs."

In the introduction to their collection on new materialism, Coole and Frost gather together a diverse range of research foci that have an "interest in the emergent materialities of contemporary coexistence" (2010: 28). These foci cover a broad spectrum of topics and disciplines: a recognition of the significance of corporeality in our understanding of culture and the human–technology relation; ways of acknowledging the causality and agency of material things; a focus on the "place" or "emplacement" of embodied beings and things in the urban, geophysical world; detailed analyses of quotidian life and our daily interactions with material objects and the environment, and their relation to "broader geopolitical and socioeconomic structures" (Coole and Frost 2010: 7); and renewed interest in phenomenological or "body-philosophy" approaches to everyday experience and perception. As Coole and Frost note, what these conceptual trajectories have in common is the view that materiality and embodiment is complex, dynamic and multiple. Indeed, as Parikka states, "the task of new materialism is to address how to think materialisms in a multiplicity in such a methodological way that enables a grounded analysis of contemporary culture" (2012: 99).

New materialism is an umbrella term for a range of research trajectories, one of which Paterson et al. and others identify as the sensory turn in the humanities and social sciences, which seek to weave together "historical, theoretical and empirical studies" (2012: 4). As Mitchell noted almost two decades ago, the "very notion of a medium and of mediation already entails some mixture of sensory, perceptual and semiotic elements" (2005: 260). Cultural analyses of contemporary mobile media thus turn our attention to how our emotional and social lives are necessarily caught up in the sensory and perceptual domains of experience, from "practices of listening" (Crawford 2012) to the use of haptic touchscreens and game interfaces (Parisi 2009). Exploring this mutual imbrication of media and our embodied selves also means delving into the intimate connection between perception and meaning that is always-already both individual and collective. Capturing this relationship between personal difference and social habitudes is perhaps one of the key challenges for researchers of media use in everyday life; as Paterson et al. argue, the "cultural chronology of the formulation of a 'sensorium' necessitates that the senses are ineluctably social: felt individually, but also always shared intersubjectively" (2012: 2).

The corporeal or sensory turn directs our attention to matters of embodiment, orientation and habituation in how we experience and interact with media technologies as part of daily life. Over the past decade, mobile media studies has enthusiastically engaged with the corporeal and embodied aspects of mobile media practices (Farman 2012, 2015; Moores 2012, 2013; Evans 2015; Richardson and Wilken 2009, 2012, 2017), and more recently in the context of mobile games (Keogh 2018; Hjorth and Richardson 2020). Concurrently, haptic media studies has emerged in tandem with developments in touchscreen and gestural devices (Parisi et al. 2017; Richardson and Hjorth 2017). All media, in these analyses, are at the same time material, experiential and affective in culturally specific ways. In the context of mobile touchscreens, it is not that such devices demand a more embodied or sensory mode of interaction, but that they have "alerted us to the sensoriality of our embodied and affective engagement with media in new ways" (Pink 2015a: 6).

Phenomenology and Postphenomenology

One of the key perspectives within the corporeal turn is phenomenology, in particular the work of French phenomenologist and body-philosopher Merleau-Ponty. For Merleau-Ponty (2004 [1962], 1964), tools and technologies are continually and dynamically incorporated as "fresh instruments" into our ways of perceiving the world. In this view, our media interfaces and corresponding user-habits work to "dilate" our being in the world. Within the phenomenological tradition, the coupling of tools and bodies is a process of *inter-corporeality*, a word that describes the fundamental interconnection between technology, bodies and ways of being and knowing.

In its phenomenological focus, our approach in this book is similarly framed within the broad premise that every human–technology relation invokes certain kinds of worlding, or mediated ways of "knowing" and "making" our collective realities. As Cooley suggests, handheld devices – always on and connected – are "productive of cultural practices that evidence new ways of being in the world," new modes of sharing and connecting, of attention and distraction, of being together and alone, of seeing and feeling, of touching and being "in touch" (Cooley 2014: 45). Perceptually, mobile media usage is quite literally a way of "having a body" that demands a complex mixture of socially agreed upon practices and sensory engagements.

Merleau-Ponty's (2004 [1962]) emphasis on our corporeal and perceptual engagement with the environment can provide valuable insights into the interpellation of bodies and tools in all human–technology relations. In particular, his notion of the *corporeal schema* or *body image* can effectively be applied to the relationship between mobile media and embodiment. The corporeal schema or body image describes the "expandable" or inherently tractable nature of embodiment. In the context of everyday activities, the experience of one's own corporeal schema is not fixed, but adapts to material and technological mediations, and cultural and historical contexts. The corporeal schema dilates in each body–technology context, as technologies and tools become literally incorporated as aspects of our sensorium:

> The blind man's stick has ceased to be an object for him, and is no longer perceived for itself; its point has become an area of sensitivity, extending the scope and active radius of touch ... In the exploration of things, the length of the stick does not enter expressly as a middle term ... There is no question here of any quick estimate or any comparison between the objective length of the stick and the objective distance of the goal to be reached ... To get used to a hat, a car or a stick is to be transplanted into them, or conversely to incorporate them into the bulk of our own body. Habit expresses our power of dilating our being in the world, or changing our existence by appropriating fresh instruments. (Merleau-Ponty 2004 [1962]: 165–6)

It is the corporeal schema that accounts for the body's capacity to intertwine with the world. Our perceptual experience is not determined by the boundaries of the material body but rather reflects the way that our corporeality extends and withdraws – changing its very reach and shape – in its dynamic apprehension of tools and things in the world. This schematic is inherently open, allowing us to incorporate technologies and equipment into our own perceptual and corporeal organization. It is this openness that affords us the capacity to use technologies as a means of extending our knowing and being of the world beyond the limits of the body.

In relation to contemporary media, it is often noted that there is a dominant visual or audio-visual paradigm that pervades our media practices, because we are largely screen-dependent, a dependency which includes our now ubiquitous mobile phones. It is worth noting here that, from a phenomenological perspective, sensory distinctions misconstrue the whole-of-body nature of perception. As Merleau-Ponty would remind us, looking, tasting, smelling and hearing are all variants of "handling" the world. Considering sensory ratios of mobile media use in terms of specific body parts, as we do in this book, is a deliberate heuristic or interpretive strategy that enables us to drill down into the details of sensory mediation, with the understanding that such practices are experienced as a complex merger of the senses which at times privileges one or more over others.

The shape-shifting of the body applies to our most banal and familiar activities. To use an everyday example, learning to drive a car involves an assimilation of the spatial organization and limits of the vehicle, its speed, the hand–wheel–direction vector and foot–brake–deceleration continuum, and so on. Within the material shape and capacities of the car, we adjust our physical deportment, spatial orientation, and our entire physical relationship with the world. Drivers need to train their car-body and accommodate entirely new ways of thinking about and moving through space: becoming familiar with emergent ratios of hand–eye and foot–eye coordination, judging distances with the visual device of the rear-view mirror (a new vision which warns "objects are closer than they appear"), and many other techno-corporeal proficiencies. In addition, we must acquire a very complex literacy of the built environment, including road width, signage and artificial lighting. Initially, learning to drive involves constant attention and concentration, because we must consciously orient our bodies towards the unfamiliar spatial and motile logic of the car; but after some practice driving becomes habitual, and our body schemas properly "dilate" to accommodate the new tool. In this way, the car becomes an aspect of our embodiment, part of our repertoire of proprioceptive skills, skills that include an awareness of the limits and boundaries of bodily influence; we have appropriated the body of the car as a temporary body or quasi-body, a supplemental way of being. Indeed, it is the essential plasticity of perception, which adapts according to a complex range of cultural, personal, technological and material affordances, that describes the very nature of embodiment. Both *as* and *in* context, our embodiment exists as a complex interaction of material and cultural environments that works to sediment body memory and habit.

While Merleau-Ponty has been critiqued for not attending sufficiently to the cultural differences of bodily experience, he did nevertheless recognize that embodiment comprises both "matter" and sociality. That is, cognitive life

> is subtended by an "intentional arc" which projects round about us our past, our future into our human setting ... which results in our being situated in all these respects. It is this intentional arc which brings about

> the unity of the senses, of intelligence, of sensibility and motility. (Merleau-Ponty 2004 [1962]: 157)

Subsequent theorists such as Ihde (1993) and Morris (2004) have further complexified this socio-material relation by including the nuances of personal practices and cultural specificity. Morris argues that the dynamics of perception "are not anchored in a fixed, objective framework, they are intrinsic to the situation of perception, and can differ across individuals, habits, and social setting" (2004: 23). Postphenomenology – a contemporary approach to phenomenology developed by Ihde – moves away from the idea that the "body" is or has some universal standard, to consider how embodiment is partially determined by sociocultural context, including gender and ethnicity, and how this inflects collective and individual sensory memories and experiences. As Ihde argues, we don't see through technologies (as if they grant transparent access to the world) but in partnership with them, so perceiving and knowing the world is entangled with the *multistability* of being human. The term multistability – as one of the central overarching concepts of postphenomenology – conveys the inherent adaptability and mutability of both bodies and technology use, always depending on the *contexts* or *situatedness* of praxis. Ihde (1993) also points out how our corporeality is always both magnified and reduced by our tools: telescopic devices, for example, may extend the reach of our vision, but while vision is extended other sense perceptions retract, turning our attention away from the immediate and tactile and towards the distant environment. In this way, as Shinkle (2003) suggests, media technologies institute material parameters, proportions of attention and inattention, by which we measure varying degrees of perceptual reach from objects and others in the world.

This book is concerned with the human–technology relation specific to bodies and mobile media, and how mobile devices "dilate" the body. Indeed, considering the number of hours that many people spend engaging with media in contemporary life, the body–screen relation in particular may be one of our most significant human–technology relations. Different kinds of content and mobile media functionalities literally change what there is to be seen, touched or heard, how things appear and what we attend to, affecting how we interact with others and how we move within and experience urban spaces and places. For example, mobile phone use becomes a way of managing the corporeal agitation of impatience, aloneness and boredom in public spaces, while at the same time we maintain an environmental knowing, or crucial peripheral awareness of our spatial surroundings, in readiness for the busyness of life to resume. The mobile device literally becomes co-opted into the corporeal labor of waiting. Such work involves micro-bodily actions as we actively and intentionally co-opt the mobile interface as means of managing attention. In this way, mobile media practices express not only a way of being in the world, but also a way of being together that requires mutual spatial and corporeal adjustment. Mobile media and network practices become enfolded inside present contexts and activities, like the embodied and itinerant acts of walking, driving, face-to-face communication and numerous other material and bodily involvements.

Following Ihde, throughout the book we consider how our contemporary engagement with mobile media is a particular kind of human–technology relation that is both stabilized and destabilized by individual and collective cultural variation. We explore such variability in terms of perceptual and experiential difference depending on contexts of use, access and bodily specificity. It is important for us to remember, as media researchers, that our modes of embodiment and perception are not uniform or neutral, but saturated with conceptual and perceptual histories, individual variation, collective habitudes and sedimented ways of being in the world and being with others. The book will include examples of diverse modes of embodiment, including disabled bodies, that will critically inform our postphenomenology. We will also discuss other marginalized and vulnerable bodies such as the elderly and those of non-Western ethnicities. Postphenomenology argues that embodiment is non-universal, so these examples will support our approach and enrich our narrative.

A Note on Metaphor

For a number of philosophers, such as Norman O. Brown (1966) and Lakoff and Johnson (1980, 1999), the bodily metaphor is the source of *all* metaphor, both as constraint and as potentiality. Metaphors are not merely conceptual or figurative linguistic devices, but essential to and formative of meaning; they are quite literally world-shapers. The embeddedness of body metaphors in daily life is investigated in detail by Lakoff and Johnson in their two collaborative works *Metaphors We Live By* (1980) and *Philosophy in the Flesh* (1999). As they argue, spatial, navigational and directional metaphors are the most common of all metaphors, and for the most part are determined by how we experience our bodies in the world: up–down, in–out, front–back, on–off, deep–shallow, central–peripheral (Lakoff and Johnson 1980: 14). These orientations are intrinsic to our motor functions and relative to our gravitational field.

The relationship between media experience and embodied metaphor is deep and sedimented (van den Boomen 2014); we use a whole range of experiential tropes to make sense of our mediated engagement with the world. In etymological terms, *trope* indicates an affinitive "turn" towards something; through body metaphors we perceptually "turn towards" media interfaces. When we use the expression "glued to the screen," for instance, we "metaphorize" our eyes as limbs entering into an intimate and tactile relationship with the interface. Screen technologies invoke a range of body tropes, such as the synecdochal "all-hands-and-eyes" experience of interactive screens and games, or the way in which the mobile phone-body becomes – metonymically – a pedestrian or vehicular node of networked communication. The metaphor of the screen as window-on-the-world, originating from Alberti's perspectival grid, remains as one of the most tenacious tropes influencing our experience and understanding of contemporary media today, evidenced by the way the user interface is commonly and metonymically taken to mean the screen interface, rendering invisible or trivial the myriad other aspects of use, such as the haptics and function of input devices, the complex sensory ratios involved in our engagement with touchscreens, or the pedestrian and locative or place-based aspects of mobile media use. Metaphors of congestion and contagion, or of movement and passage through space, are often adopted as explanatory tropes for the transmission and corruption of digital media and data. We use spatial models of touring, mapping, topology and geometry to "locate" ourselves *in* media and mediated spaces.

As we discuss in Chapter 4, hands and fingers take on a familiar metonymic function as they "stand in" for the perceptual body. With mobile media there is a fundamental and irreducible relation between bodily dexterity (knowledge in the hands and fingers) and our habits of visual and embodied perceptual orientation and movement; we use numerous and familiar body-based metaphors to conceptually and perceptually manage our engagement with mobile interfaces – "scrolling" through content on Instagram, "mapping" our location on Google Maps, "finger-bombing" pigs in *Angry Birds*, "navigating" web browsers as movement forwards and backwards, experiencing elasticity and gravity in mimetic mobile games, or perceiving "things" in virtual spaces (like trash bins, treasure chests, or even the metaverse) as types of containers. We also see culturally specific embodied metaphors present in the history of mobile phones, where, for example, portability is captured in the Japanese term for a mobile phone (*keitai* = portable) and tactility and handiness is captured in the German word (*handy* = mobile phone).

The robustness of embodied metaphor is particularly evident in our corporeal and conceptual familiarity with the experience of telepresence – perception and communication at-a-distance. The intimate and adaptable connection between media and sensoria effectively works to narrow communicative distance by enabling us to redefine what it means to be present. Technological developments ranging from the telephone through to radio, television, cinema and video games have created mediated spaces where a sense of presence can be felt beyond the location of the physical body. More recently, online and networked media have irretrievably altered our sense of embodied location and presence such that it is no longer possible to definitively distinguish between previously oppositional categories of here and there, near and far, private and public space, dialectics which dominated our understanding at the beginning of the twentieth century. The mobile phone, in particular, works to enfold contexts, such that urban spaces are now filled with mobile phone users who create communicative pockets of co-existing modalities of co-presence – telepresence, absent presence, distributed presence and ambient presence – all of which deploy embodied metaphors of tactile, aural and visual access as a means of altering our experience of intimacy and being together.

Presence and intimacy are "felt" differently by each of the senses, and we have different corresponding metaphors in each case, exemplified by familiar phrases such as "I'll be in touch" (as a metaphor for communicative presence), "I hear you" (an expression of understanding and "being there" for someone), "take a walk in my shoes" (conveying empathic experience), "shed some light on the matter" (making visible an explanation) or "face the music" (confronting responsibility). As we discuss throughout the book, such metaphors are part of – but also transformed by – our perception and understanding of mobile media practices. The "sensing" of mobile media, an intimately audio, visual, haptic and sometimes visceral awareness, enables a mode of embodiment and knowing of the world such that experiences of distance and closeness, presence and absence, are no longer simply by their definition at odds.

Our ability to embrace mediated perception and modes of embodiment within our body schema – and to oscillate between, conflate or adapt to ostensibly disparate modes of being and perceiving, by way of embodied metaphor – is precisely why telepresence or distant presence is perceptually tolerable and available through all our senses. As we suggest, we call upon an "as-if" structure of presence and mobility – enacted through tropes of presence and movement – that is fundamental to our experience of networked and mobile media.

We argue throughout the book that such bodily tropes and metaphors play an essential role in our understanding and experience of mobile media interfaces and practices. Embodied metaphors allow us to translate our perception and experience of mobile media as integral to our lived realities; the "taps, pushes and sweeps" of mobile touchscreens quite literally "challenge the familiar concretia of the world" (Elo 2012: 5), as evidenced by the way we endow media objects with skeuomorphic real-world effects and actions. Examining the tropology of mobile media provides a valuable way of understanding our embodied engagement with mobile devices, the corporeal turn in scholarship more broadly, and the significance of the media–body and mobile–body relation in contemporary life.

Chapter Overview

Each chapter will emphasize a specific sensory modality as well as examine how each modality is embedded in multi-sensory experience and histories of media practice. As already noted, focusing on specific senses allows us to articulate and critique how there are sensory hierarchies at work in our use of media, consider how this is historically sedimented in our experience of media interfaces more broadly, and examine how different dimensions or modalities of use involve specific sensory affordances. The final chapter takes a different approach by exploring the corporeality of mobile media more holistically as a synesthetic experience.

Although high-level theoretical arguments inform the conceptual framework and arguments throughout the book, our aim is to translate postphenomenological analyses of mobile media practices, and the corporeal turn in scholarship more generally, in accessible ways for a wider audience, through detailed examples and engaging anecdotes that convey the experiential and sensory feel of media experience. While the book will be focused on mobile media it will also reference and engage with other media forms and screen experiences.

In the next chapter we consider how our "facial" engagement with screen interfaces can be understood in terms of embodied metaphors. We explore the face–screen relation, and suggest there is a certain material and physical affinity between faces, windows, frames and screens, in terms of the way we "turn" to them with varying degrees of attention and distraction. In the context of mobile devices, the frontal or face-to-face relationship we have with screens is modified to adjust to our own macro-mobility (i.e. walking) and micro-mobility (gestural engagement with the touchscreen) and the interrupted nature of mobile phone use. In a very fundamental way the mobile interface modifies what we pay attention to, what we "turn to" and face (and turn away from) in the everyday lifeworld, and the modalities and *durée* of that attentiveness.

As Robert Romanyshyn insightfully observed over thirty years ago, the primacy of vision has become "a habit of the mind" – so many of our technologies and media interfaces "emphasise this feature of visibility ... we might venture to say that our sense of reality has nearly become identical with our ability to render something visible" (Romanyshyn 1989: 184). In Chapter 2 we trace the history of ocularcentrism in the context of media screen practices, and consider the role of vision in the eyes–hands–screen circuit that typifies our use of mobile media, providing a range of examples from *Pokémon GO* augmented reality gameplay to mobile phone photography. The discussion will also revisit our experience of Google Glass and Spectacles by Snapchat in terms of how such eyewear complicates the cultural contexts of the "look."

In *Understanding Media*, Marshall McLuhan famously suggested that we should consider the telephone as an extension of the ear (McLuhan 2003 [1964]: 289). Taking up this suggestion, in Chapter 3 we explore the ear–mobile device coupling, and the particular sensory modality of hearing. We first trace the histories of mobile device use and practices of listening and aural/oral connection, giving special attention to the complex relation between hearing and temporality. We then examine typical scenarios of use, such as the "immediacy" of interpersonal conversations (across Bluetooth, speakerphones, etc.), the uptake of casual mobile games as "fillers" in interstitial moments of waiting, the "out-of-time" privatization of listening and the "cocooning" of the self via music and podcasts, and the mobile medium itself as a "listening" device that captures individual micro-data in our collective shift to the "quantified self" that co-opts time as a resource.

In Chapter 4, we consider hands and tactility. Hands and touch – the tactile sense – are central to our contemporary engagements with mobile devices. We hold mobile phones in our hands and operate and interface with them using an entirely new gestural literacy. This chapter is framed around an examination of haptics and of the hand as a world-shaping tool. In it, we trace the emergence and significance of haptics, exploring histories of tactility (Parisi 2018; Paterson 2007), and the acceleration of haptics with the arrival of touchscreens. In thinking about our embodied engagements with haptics, we also examine the gestural economies that accompany and have emerged around touchscreen use, and we explore how there is a naïve physics to touchscreen interfaces that means they function as only a loose simulation of real-world actions. In the latter part of the chapter we consider if and how mobile media might incompletely "stand in" for social bodily touch, as it did during the recent pandemic, by way of our ability to enact an "as-if" structure of perception.

Walking – the focus of Chapter 5 – is widely understood as fundamental to our corporeality, ontology and cultural practices. Being "on foot" is argued to be amongst "our most elemental senses of our standing in this world"; walking establishes "intimate contact" with places and makes them "dense and prickly with details and complexities" (Amato 2004: 277, 276). Such is the richness of the experience of walking, Tim Ingold argues, that it can be seen to constitute "a form of circumambulatory knowing" (2004: 330). In this chapter, we examine how pedestrian mobile media use, explored across a number of different examples, impacts upon everyday practices and pedestrian movement in public urban spaces, thereby modifying our ways of walking through the city. Our argument is that walking with and while using mobile devices requires particular kinds of body-work, and particular kinds of body–technology and body–place relations. The peripatetic modality specific to much urban mobile device use involves a situated, corporealized and often personalized negotiation of place, space, co-presence and environmental "knowing," enacted through the relational engagements between the pedestrian body and the mobile screen. Returning to the earlier discussion of the face and eyes, we also examine how the downturned gaze of the mobile phone pedestrian is said to disrupt the facial cues for navigation famously noted by William H. Whyte (1980) in his study of urban pedestrian movement.

The Conclusion will weave together the sensory strands of mediated experience traced throughout the book. That is to say, in addition to providing a summation of the preceding content and arguments, this final chapter also offers a more holistic interpretation of mobile media practices as multi-sensory and synesthetic – moving beyond an account of specific sensory ratios or body–technology couplings to detail how our sensory modalities operate in combination to create more fully embodied technology relations and engagements with mobile devices. In the final section of this chapter, we address those sensory modes of engagement not given specific chapter-level attention – namely, taste and the olfactory sense – using these as a point of entry for thinking about emerging work on corporeality and body–technology couplings, especially as they relate to embodied, multispecies engagement with screen media (Richardson et al. 2017; Webber et al. 2017).

1
Face

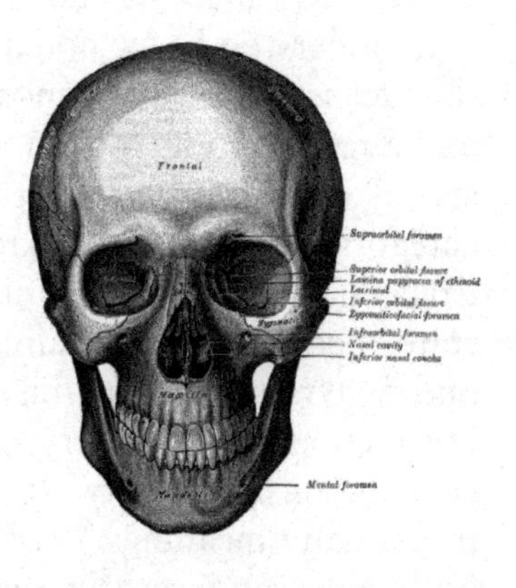

For human beings, faces have always been a privileged site of visual stimuli. We constantly observe and monitor each other's faces, assessing them in accordance with cultural, social, and increasingly, technological constructs. (Sumner 2022: 1)

The human face consists of the front of the head extending from forehead to chin, featuring the nose, eyes, mouth and cheeks, and is the means by which many if not most of our emotions are expressed. It is comprised of between twelve and seventeen bones (depending on how they are categorized), and twenty craniofacial striated muscles that control the two major tasks of chewing and facial expression. With its combination of anatomical and functional structures – housing the three major sensory organs of sight, smell and taste – the face is also highly variable in terms of different qualities (e.g. color of the skin, shape of the eyes and nose) and quantities (length of the jaw, size of the forehead). In zoological terms, the incredible variety in our faces – much more than is found in other bodily traits – is due to the way a part of our brain is specialized to recognize these differences, as distinct from other animals who identify individuality via smell or vocalization (Sanders 2014). Despite this variation in form, what constitutes the human face is universally shared and understood. As noted by Sumner (2022) above, the face is arguably the most predominant loci of visual and affective communication – a "dynamic information space" and "transmitter of multiple and complex social categories" – conveying external and internal characteristics such as gender, ethnicity, age, health, emotional state, personality, pleasure, pain and deception (Jack and Schyns 2015). In this respect, we might consider the face as the originary or primordial interface. The conceptual complexity of the word "face" is reflected in its many meanings – with over twenty definitions as both noun and verb in the English language, ranging from the biological and material (the face of the earth, the human face) to the figurative and metaphorical (pulling a face, at face value, a more human face).

This chapter considers the complex relationship that exists between screens and faces in day-to-day life – from the televisual to the many computer and mobile screens encountered in both domestic and public spaces – and suggests that each of these encounters has its own corporeal and "interfacial" modality. More specifically, the discussion will explore the relational and frontal ontologies of the face and the screen interface, focusing on the specific body–technology relations that emerge from our bodily incorporation of computers and mobile screens. In particular, we will suggest that our engagement with media screens at a perceptual and embodied level can be understood by way of phenomenology, supplemented by a consideration of the various ontological tropes and "body metaphors" that are deeply embedded in our experience of screen interfaces. This focus on the perceptual and metaphorical aspects of the body–screen – and more specifically, face–screen – relation, can provide some insights into the historical and material affinity between faces, windows, frames and screens, and the complex ways we "turn" to them with varying degrees of attention and distraction. Finally, we aim to show how this affinity is challenged at a fundamental perceptual level by our experience of contemporary mobile screens.

Medium Specificity

As we have previously noted, our approach is framed within the broad premise that every merger between our bodies and technology involves certain kinds of being-in-the-world and particular ways of knowing and making that world. In these terms, the screen interface is quite literally an aspect of our individual and collective corporeal schemas; that is, through routine use screens have become part of the dynamic arrangement of our embodied experience. A corollary to this approach is the notion that our engagement with screens and interfaces is medium specific, such that each screen modality – whether televisual, computer or mobile – effects a different mode of embodiment, a different way of "having a body." Yet the incorporation of screens into our corporeal schema is also determined in part by cultural, environmental, spatial and historical specificities – by the habitudes of practice that have developed within the contextures of everyday life. Thus, for example, as television theorists such as Morley have argued, "television" and "home" have redefined each other (Morley, in Jenks 1995). Early conventional television architecturally transformed the living area of the home into a "viewing space," requiring modifications to how the body was habitually positioned and mobilized, while over the past decade the proliferation of televisual entertainment technologies within the home has effected new ways of carpentering the built environment, to literally "make room" for new media spaces, by way of open-plan design or the designation of a cinematic niche for the home theatre or entertainment center. As media theorists Silverstone and Hirsch suggest, the actual location of the TV set has implications for our embodied and spatial experience of both the interface and the immediate environment, including our placement and proximity among other viewers and domestic objects. Such studies have shown that the screens dynamically *transform* the environment of reception – quite literally what we pay attention to or "face" in the space of the home – and the embodied experience of domestic media practices.

More recently, mobile phone research has provided deep ethnographic and comparative analyses of mobile media cultures and practices in the home and in urban spaces, and of the way mobile devices have personalized and fragmented media experiences while we are sedentary or on-the-move (Hjorth and Richardson 2020). While televisual interfaces invited us to face the screen together and consume the same content (hence the term "electronic hearth"), critics such as Turkle (2012) argue that individualized technologies such as mobile phones now capture our separate attention and turn our faces downwards and away from communal experience – whether we're sitting "alone together" in the shared space of the living room, or on the train commuting to work. Yet as ethnographic observation reveals, the situation is considerably more complex and multistable. People have diverse and unique media screen practices, across digital, networked and face-to-face contexts, each of which impact on the configuration of domestic and urban spaces and the facial attention given to our screens and each other. For one family with three children, for example, the family room buzzes with activity. The children share a close bond through *Minecraft*, with the eldest on her laptop and mobile phone chatting to school friends on Discord and WeChat, and the younger ones on their tablets. The family's couch is now a makeshift workstation for the eldest child, and the coffee table a gaming hub for the younger children. The children often take breaks from their screens to show off their *Minecraft* creations to each other and their parents, sharing their impressive structures or favorite locations in the game. At the same time, they talk with friends online about school, friends, family, pets and their gameplay. The family room, once a space for watching television, has transformed into a space for cross-platform digital entertainment through multiple devices. Far from being "glued" to their mobile media or computer screens, the interfaces are intimately woven into the "rhythms and routines" (Hjorth et al. 2020) of everyday life in ways that seamlessly merge online and offline interaction.

Ihde argues that the body–technology relation is our fundamental ontological condition, yet each of the relations that define and transform our techno-perceptual experience is *non-neutral*, specific to a myriad of contexts. Mobile screen practices, like all such relations, are inherently multistable. As Choi (2007) noted, drawing on the work of Ting-Toomey and Kurogi, there are significant cultural disparities in the performance of self – and the corresponding "face-negotiation" strategies required – that become embedded in mobile media cultures and mobile phone practices. Thus, for example, in Japan, there is an explicit distinction between *honne* – "true feelings" that are kept to oneself – and *tatemae* – one's public face; while in Korea, *nunchi* – the ability to "read" and interpret others' faces and social cues – is regarded as an essential skill. Choi argues that such face-negotiations and customs (among other technocultural specificities) are intrinsic to mobile phone use, revealed by both the subtle and more palpable cultural differences evident in the everyday "work" of image sharing, social media practices, texting and gaming. Such disparities in the micro-practices of mobile phone embodiment reveal how both collective and personalized mobile media practices are in fact a complex coalescence of cultural, cognitive, material and somatic factors.

In his phenomenological analysis of our prolific visualizing technologies – from domestic, personal and public screens to highly sophisticated scientific apparatuses – Ihde documents how the body and instrument form a temporary collusive entity that handles the world in specific ways. In each case, he writes, the mediation must be made to "fit" the body, and, in particular, there exists a consonance between the device and our "face-to-face capacities":

> The mediated presence ... must fit, be made close to my actual body position and sight ... What is seen must be seen from or within my visual field, from the apparent distance in which discrimination can occur regarding depth, etc., just as in face-to-face relations. But the range of what can be brought into this proximity is transformed by means of the instrument. (Ihde 1990: 72)

Ihde's analysis allows us to consider the ways in which different media and screen interfaces effect different kinds of perceptual and communicative reach, though for the most part our experience of and orientation to the interface is determined by the need to see and therefore "face" it. As we will suggest, however, while our perceptual engagement with screen interfaces is often predicated on this face-to-face configuration, contemporary mobile screens frequently work to confound or at least problematize this relation.

As noted, this distinction between various body–screen modalities explicitly acknowledges the concept of medium specificity, a term originating from the work of technological determinists Marshall McLuhan (2003 [1964]) and Harold Innis (1964). Medium specificity describes the fact that specific media have specific spatial, temporal and sociocultural effects, determining particular conditions of possibility for the way meaning is made. While Innis was concerned with the historical breadth and evolving political effects of communications technologies on cultural formation and social organization, McLuhan claimed that all media are extensions of the body: they alter our sensory access to the world, determining and organizing our experience, our forms of knowledge, indeed the very structure of perception. In McLuhan's understanding of medium specificity, each communication medium works to "fix" particular sensory ratios, stipulating forms of knowledge and orchestrating the structure of perception by "attuning" our sensory equipment to absorb reality in medium-specific ways (Carey 1969: 284). While acknowledging McLuhan's insights, beyond the idea of "sensory ratio" we would describe the screen–body coupling in more relational terms to include the way in which the body–media relation is also moored by sedimented cultural habits, body metaphors and tropes surrounding our engagement with screens, and the impact of the situated or built environment upon that engagement.

Thus, for example, we often refer to the difference between our engagement with conventional broadcast or streaming televisual screens and interactive computer screens in terms of how we choose to position the body when attending to the screen; that is, when watching television we "lean back" in contrast to the "lean forward" body posture demanded of interactive screen media, where there is an imperative to face the screen more proximally and directly. This describes the variable embodied and facial orientation we have towards different kinds of media interfaces, and the immersive investment of the eyes, ears and hands needed for interactive screens. In a fascinating study of the "neck-down" posture required of face-to-screen handheld mobile media consumption, Sarria et al. (2022) document how consequent pressure on the spine is caused by spinal flexion that restricts "typical" expansive human posture and demands more exertion (up to sixty additional pounds of force). For many people this results in the physiological stress of "tech neck," and can be linked to lower attentional engagement and message processing, and negative effects on cognition, emotion, confidence and pain tolerance. Sarria et al. argue that the results from their study showing differences in response across postures "suggest that communication theories should be revisited and reinterpreted given the different postural world we live in when it comes to media consumption." Such research reveals that the relation between screens and bodies, and the spatial arrangements, dimensions, functionality and interfacial specificities of such screens – including one's mode of bodily involvement and facial posturing – at least partially determine our degrees of attention, cognitive acuity and emotional attitudes of reception.

Embodied Metaphor: Facing the Screen

In what follows, we will further explore our bodies' facial (and interfacial) involvement with screen media as quite literally *mediatropic*, suggesting that both body and screen are imbricated in a number of complex ontological and embodiment metaphors. If we remember that the combining form *-trope* indicates an affinitive turn towards something, then screen interfaces can be said to have had significant "tropological" effects on our bodies; our modes of embodiment "turn towards" specific technologies and media interfaces. Indeed, the explicit goal of media designers in general is to render the screen "sticky" as a measure of viewer adhesion.

Lakoff and Johnson claim that embodiment and material metaphors are embedded in all our experiences. They categorize these metaphors as ontological metaphors, or more specifically as entity, substance and container metaphors. They write:

> We experience ourselves as entities, separate from the rest of the world as containers with an inside and an outside. We also experience things external to us as entities often also as containers with insides and outsides. We experience ourselves as being made up of substances, e.g., flesh and bone, and external objects as being made up of various kind of substances, such as wood, stone, metal, etc. (Lakoff and Johnson 1980: 58)

In *The Production of Space*, Lefebvre identifies the crucial work of metaphor upon the body, suggesting that metaphors are not simply figures of speech, but rather decipher the world into that which is "sayable" or "susceptible to figuration"; in so doing, acts of metaphorization take as their point of departure a "body metamorphosed" (Lefebvre 1974: 139–40). Thus, all bodies are caught within a complex web of analogies and conceptual metaphors. Metaphors, then, are the extension of our corporeality into the world: only that which can be *metaphorized qua embodiment* – interpreted in terms of our complex body-model – is realized or *made real*. We literally and figuratively project our bodies in-the-world. The face and eyes are body parts that we most commonly map onto other things and concepts. The facial metaphor, for example, is used to refer to that which both contains and displays our emotions (e.g. a straight face, anger flashing across the face, making a face), to the surface of a concrete thing (the six faces of a dice, a watch face, the face of a cliff), to a conceptual external image of an abstract domain (the public face of the company, putting on a brave face), or in a metonymical sense to stand in for the whole person or thing (showing one's face as a trope for identity).

For Lakoff and Johnson, humans (and animals in general) have a front and a back, or a face and behind, and we embed this ontology or understanding of being-in-the-world into the constitution of spaces and objects in our worldly environment (1999: 34). There are many instances of this "frontal ontology" in our use of technologies and the way in which we navigate them. For example, the standard graphic user interface on a computer or mobile screen is configured in such a way that we experience our progression through directories in terms of forward and back, in and out, up and down. These common navigational and browser spatialities, along with other body metaphors adapted to virtual spaces, are clearly and quite simply based in our bodies' engagement with the world. Importantly, these schemas are not just outcomes of physiology, they are also culturally specific, and vary from culture to culture. Culturally specific body-orientations are often instilled at a very young age; for example, in contrast to the dominant Western habit of facing a newborn baby towards the holder, "Kaluli mothers tend to face babies outwards so that they can be seen by and see others that are part of the social group," habituating a particular orientation to both the maternal and wider environment (Woodhead et al., cited in Donald and Richardson 2002). In a more general sense, as humans we project fronts and backs onto things, and habitually designate the "face" or "front" of many objects as the aspect with which we interact, because we ourselves face them. As Lakoff and Johnson note:

> The concepts *front* and *back* are body-based. They make sense only for beings with fronts and backs. If all beings on this planet were uniform stationary spheres floating in some medium and perceiving equally in all directions, they would have no concepts of *front* and *back*. But we are not like this at all. Our bodies are symmetric in some ways and not in others. We have faces and move in the direction in which we see. Our bodies define a set of fundamental spatial orientations that we use not only in orienting ourselves, but in perceiving the relationship of one object to another. (Lakoff and Johnson 1999: 34)

Clearly, most of our communication technologies are oriented in this way; moreover, even when their purpose is not to provide visual images, they more often than not still have "faces" from which we read information displays. While there is no doubt that we have a primarily "frontal" relationship with the screen, this is not to say that we have no association with the "backs" of such devices, although these interactions are for the most part brief and functional, that is, for the purpose of connection, or negotiating an effective relationship with the front. We thus have an affinity with the body of the screen simply by virtue of the fact that human bodies and screens have "fronts" and "backs" and "face" each other. It is this screen–face consonance which perhaps best explains the phenomenon of parasociality, and the common behavior of reacting to media content "as if" the latter represents real people and real places. The social and behavioral aspects of this affinity between humans and media screens is explored in some detail by Reeves and Nass (1996) in their much-cited study *The Media Equation*.

As noted, this front-to-front relationship is one that we have with screens in general. In most if not all cases, the screen is a frame of limited dimensions within our own physical space, while the body's frontal relationship with the apparatus varies between media, depending on what Manovich calls "viewing regimes" (2001: 96). With cinema, for example, the viewer is at the outset fully frontal to the exclusion of all diversions, focusing entirely on the screen. In the optimal situation, the boundary or interface between body and cinematic apparatus dissolves, manifesting a change in orientation from being "in front of" to being "within," an effect which is achieved by several factors: the size of the screen, the darkness of the theater, and not least the surround sound. Front-to-front orientations are therefore not achieved by vision alone; in many situations, when facing a moving image, we would expect that sound would also approach us from this direction, but the effect of surround or stereophonic sound is to embrace the body in such a way that the frontal relationship with the screen is at least partially compromised. In the case of television – with perhaps the home theater an exception – the face-to-face relationship between the body and the set is somewhat more informal and less disciplined: viewers can look away to the familiarity of their domestic surroundings, move about or leave the room; they can be visually and aurally attentive or inattentive to varying degrees, by muting the sound, zapping through channels, talking on the phone or conversing with co-watchers, reading or engaging in other activities. In other words, the facial and sensory dedication we apply to media screens varies according to the mode of techno-somatic involvement demanded by both the interface and the cultural, experiential and material contextures.

Although in a general sense we clearly have a frontal and gravitational ontology that impacts the way in which we perceive and navigate screens, the emergent body–media relation we have with mobile screens has seen a number of adjustments to this corporeal schema; for example, the various postures surrounding mobile phone photography, the practice of sharing one's screen with others, or more simply developing habitual skills, such as becoming adept at texting and scrolling while walking. In these cases, the often-dedicated frontal orientation we have towards larger screens becomes compromised both by our own mobility, the size and resolution of the screen, and the interrupted nature of mobile phone use. In their early study of mobile video communication in everyday life, O'Hara, Black and Lipson (2006) examined the medium specificity of video phoning, revealing that a different set of somatic adjustments is needed.

Of most interest in O'Hara et al.'s study is the ergonomic incompatibility between moving bodies and video communication, and the often-uncomfortable fit between facial and visual attention, voice/video communication, macro-mobility (walking) and micro-mobility (adjusting the position and orientation of the phone). For example, using the videophone feature requires a return to a more visually determined face-to-face orientation with the screen; that is, holding the phone out at arm's length with the screen directly – and fixedly – in front of the face. This necessitates use of the speaker phone, such that both the screen display and the usually private voice communication becomes public (O'Hara et al. 2006: 875–6). In some instances, this means that the proper boundaries between public and private cannot be maintained – both in terms of intruding voices and images into another's personal space (on a bus, for example), and in terms of exposing *both* sides of one's own private communication by varying the customary (aural) and intimate somatic mode of mobile phone communication. O'Hara et al. note that while recipients of a video call could put the phone down and use the hands-free speaker, "it was considered rude" to create a visual asymmetry between speakers such that they were no longer communicating screenface-to-screenface (2006: 878). For these reasons, participants in their study used video phoning only in particular situations – for the most part, video calls were made when the phone would be shared amongst a group of friends, or for consolidating "special relationships" that required dedicated face-to-face time even when not co-present. In other words, unlike the casual brevity of the text message or the spontaneity of voice calls, video phoning is designated for calls of some import that require a more *deliberate* attentiveness to one's bodily commitment and involvement, and the displaying of one's face to another. During the Covid pandemic, which saw a rapid increase in video communication – on both computers and mobile phones – Sumner (2022: 9) wrote that "we must reconstruct our learned frameworks of perception to interpret the gestures, facial expressions, and other movements" as the magnified and metonymic interaction with others' facial images alters the sense of physical distance we normally experience in co-located encounters.

The Window on the World

Another interesting entry point into the body– and face–screen relation is to consider one of the more common metaphors of the screen – that of the frame or "window-on-the-world." The ontological and cultural significance of the window and the frame cannot be overstated; as Friedberg comments, in her important exploration *The Virtual Window*, the frame is perceived as "the decisive structure of what is at stake" in contemporary media experience (2006: 14), while for Sobchack it is both a "lived logic" and itself "an organ of perception" (Sobchack 1992: 134). The comparison between screen and window as framing devices is easily made and understood – the frames of window and screen are similarly rectangular, they can be similarly interpreted as membranes between "inside" and "outside," and what one sees through the frame is a portion of the world in space and time. It is worth examining in some detail the portrayal of the screen as frame or window, and how such a rendering clearly instantiates a particular kind of relationship to the body, its orientation and its facial involvement with/in the medium.

The window-on-the-world is a trope emergent from linear perspective. In the space of linear perspective, the observer looks at the world as if through a window. The "tropological effect" of linear perspectival vision and the window-on-the-world can be characterized by the way visibility and light have come to stand for truth, belief and knowability. The corporeal effect here is clearly one which elevates visual perception and the eyes as that which can most accurately deliver the truth of something. As Romanyshyn argues, this put the hegemony of the eye firmly in place, such that "Alberti's window, which begins as an artistic device, thus becomes a style of thought, a cultural perception, a way of imagining the world. The window as membrane becomes the boundary, the place where the world is divided into exterior and interior domains" (Romanyshyn 1989: 69). Romanyshyn insists, then, that the window of perspectival vision set up an ontological boundary and distance between the space of the observer and the space of the observed. Significantly, in the case of Alberti's window – an artist's technique famously developed by Leon Alberti (c. 1435) for accurately capturing perspective by placing a uniform grid between the scene and the viewer – bodily movement is restricted or even absented by the device, in that the grid needs to remain directly between the scene/seen and the line of sight: the body is at the service of vision and facial orientation.

In phenomenological terms, we might consider the window, frame and screen as perceptually interfamiliar, exhibiting a kind of experiential consonance; as Friedberg notes, like the window, the screen with its frame "holds a view in place," becoming a transformative aperture in architectural space, altering the materiality of our built environment and opening surfaces up to a new kind of conceptual and metaphoric "ventilation" (Friedberg 2006: 1). The screen-as-window, then, sets up a particular kind of corporeal trope: to look out a window and to view a screen, at the imperative of the eyes and face, one's body must be turned towards the apparatus. As such, because of how our eyes are positioned on the face, to remain visually attached the body is rendered immobile. Indeed, for Manovich, this fixedness typifies a bodily inertia and sensory deprivation that has been a predisposition of "the Western screen-based apparatus" in general (Manovich 2001: 104). This tendency can be traced from Alberti's perspectival window and Renaissance monocular perspective, through to Kepler's *camera obscura*, the nineteenth-century *camera lucida* and contemporary cinema: in all these interfacial experiences, the body is fixed in space. Although the dynamic screens of cinema and television might be said to virtually or sensorially transport the viewer, Manovich argues that this mobility comes at the cost of the "institutionalized immobility" of the body of the spectator (2001: 107), in the form of the silent seated rows of movie-goers or the domestic couch-reclining TV viewer. Interestingly, Manovich claims that this condition of the body's immobility can also be traced through the history of communication:

> In ancient Greece, communication was understood as an oral dialogue between people. It was also assumed that physical movement stimulated dialogue and the process of thinking. In the Middle Ages, a shift occurred from dialogue between subjects to communication between a subject and an information storage device, that is, a book. A medieval book chained to a table can be considered a precursor to the screen that "fixes" its subject in space. (Manovich 2001: 104–5, note 48)

The mobile phone and video phone, mobilizing the communicator according to the imperatives of push media and perpetual connectivity, are devices that perhaps return us to the act of walking-and-talking, as exemplified in the practice of "feeting" (walking meetings) popularized during the Covid pandemic.

The rectilinear dimensions of the media window – and its immobilization of the body – are an instance of the epistemological containment of knowledge in perspectival vision, today most familiar through the ubiquitous frame of the screen. Thus, by tracing a lineage from Alberti's window to contemporary screen technologies such as television and cinema, we can see the *medium specificity* of our understanding, our spatial and somatic perception, and what is often termed our frontal ontology.

There is no doubt that, in some aspects, the mobile screen and the window are similar. Both are surfaces in focal disappearance – they are looked through or past rather than at, such that the material "glassiness" of their surface recedes from awareness when the window is clean and closed, or when the mobile screen is on. Yet corporeally and phenomenologically their differences are perhaps more significant. The window can be opened, shouted through, become stuck, or objects can be thrown *through* it both open and closed; in the latter case, the window can be broken in such a way that the mobile screen cannot, for the visual field remains. Moreover, it can be "handled" as a boundary to keep out the world – rain, dust, noise, people, etc. – and often works as a secure limit between domestic/inside and public/outside space. The window is thus caught up in quite distinct body–tool relations. The mobile screen, on the other hand, has a technical "body" which can be moved and carried about from one place to another, a liminal window that remains fixed in the structural design of its housing. In this sense, the mobile device as physical object – as material shape, substance and electronic interiority – enters *into* the lifeworld, and unlike the window attaches to, and travels with, other bodies and objects there, altering the spaces and habits of everyday life.

While it is the case that the frontal or frame-ontology of windowed perception remains as one of the most persistent embodied metaphors still influencing our engagement with contemporary media (despite the fluctuating successes of virtual reality technology), there is no doubt that this trope is challenged by the haptic and facial relations that have emerged from our use of smaller portable screens. Even our embodied relation with the personal computer screen is quite different to our experience of both traditional televisual and cinematic screens in terms of proximity, orientation and mobility, and not least because we are no longer "lean-back" spectators or observers but "lean-forward" users. In particular, our face-to-face relation to the computer and mobile screen is intimate, up close, and involves the negotiation and manipulation of a networked screen-space via the keyboard, mouse or touchscreen, setting up an interactive circuit of eyes, ears, hands and interface. Moreover, laptops and handhelds can be carried with us, in our hands, pockets or bags or on our laps, effectively mobilizing the body– and face–screen relations into the workspace, pedestrian space, vehicular space and the numerous public spaces of the urban environment.

The vacillating degree of attention and distraction particular to mobile screen use also problematizes the frame ontology and the facially determined body posture proper to the window metaphor. In what follows we explore this departure further in terms of the oscillating registers of attention, inattention and distraction enacted when engaging with small mobile screens, and suggest that such engagement undermines both the facial dedication of the immobilized body deemed typical of our embodied relation to larger screens, and consequently the frontal ontology of window and frame.

In considering the registers of attention and distraction particular to the screen–body relation, it is useful to identify some general ontological or material properties of what Introna and Ilharco called "screenness," addressing the questions: what is a screen, and what does it afford? In contemporary life screens are often a primary focus of our attention and concern: they literally display that which is relevant or worthy of notice. This property of relevance has little to do with the specific content of any particular screen display; it rather indicates:

> a particular involvement in-the-world in which we dwell and within which screens come to be screens. It is not up to anyone of us to decide on the already presumed relevance of screens; that is what a screen is – a framing of relevance, a call for attention, a making apparent of a way of living. (Introna and Ilharco 2004: 227)

Introna and Ilharco suggest that screens of all kinds enter our involvement-in-the-world at the moment we turn them on, at which point we reposition our attention and "sit down, quit – physically or cognitively – other activities we may have been performing, and watch the screen" (2004: 225). Yet this "frontal" relationship, typical of our engagement with most screens – where the mediums of cinema, television and desktop computers can be said to discipline the body more or less into a face-to-face interaction – is thoroughly disintegrated by the mobile media screen. Our interaction with mobile screens is rarely marked by such dedicated attentiveness; indeed, our "turning towards" them is often momentary (checking for a text or missed call) or measured in minutes.

Mobile media also elicit variable levels of attention and inattention that shift between actual and telepresent space, partially depending on the demands of the immediate environment and the extent to which the interface becomes ready-to-hand in a Heideggerian sense (i.e. its function and usability become intuitive or part of our body-memory and recede from explicit awareness). Thus one's own body may "behave" in ways that accord with (or deviate from) consensual and recognized modes of being-on-the-phone, such as stopping, bowing the head to conceal the face and reduce audibility, shielding one's mouth with the hand to define a provisional private space, or deliberately not altering one's trajectory or visual/facial orientation, and directing one's gaze into the middle distance, as is the case with the Bluetooth pedestrian. To borrow from Goffman's useful analysis of pedestrian traffic, in such responses the mobile phone pedestrian articulates a specific and recognized type of "gestural prefigurement" or "body-gloss," which intentionally displays to others a state of being-on-the-phone (1967: 31–2). The various postures and embodied actions particular to mobile phone use in public places, and the accompanying dynamics of attention–inattention, are quite specific to the body–mobile relation which has emerged over the past three decades. Here, the typical "phone-face" we customarily adopt when on the phone (eyes looking into the middle distance, with attention focused on the interiority of one's aural sensory perception) becomes also a public face with which the gestural body is aligned, a face-and-body that says, "I'm on the phone." Similarly, the activity of casual gaming or noodling with one's mobile media device while waiting for a friend or at a bus stop becomes another way of managing one's alone-ness in public spaces, enacting a particular kind of "face-work" in Goffman's sense. The transient and non-dedicated attentiveness required by the small screen and the casual game – you can "switch off" but "not totally" – allows the user to avert their gaze from others and so cooperate in the tacit social agreement of non-interaction among strangers. Here, the micro-mobilities of the body quite literally enact a mobile-specific *mediatrope* – inclined metaphorically, corporeally, communicatively and gesturally towards the mobile media device.

Face to Interface: The Selfie

There are many examples of the ways that mobile media screens challenge conventional screen–body and screen–face relations, including image-sharing practices, location-based and casual gaming, livestreaming, the practice of posting to social media while on the move, or simply the more mundane activities of talking and texting. Indeed, if each new mobile media functionality can be considered in Merleau-Ponty's terms a "fresh instrument" which dilates our corporeal being accordingly, then we are continually in the process of learning a new range of collective bodily skills, spatial perceptions, postures and habits. Although in a general sense we may have a frontal and gravitational ontology that impacts upon the way in which we perceive and navigate screens, the emergent body–technology relation we have with mobile screens has seen adjustments to this corporeal schematic. Mobile phone photography, for instance, could be said to have impacted the nature of face-to-face communication and social intimacy across both screen and co-present interactions.

Today, camera functionality has rendered the face a visual password that permits entry to one's own mobile device. The mobile camera phone has also altered everyday photographic practices in terms of perpetual photo-readiness, enabling the capture of immediate and often intimate objects and events. Users will share photos and video just taken or received with others in face-to-face and mediated interaction, by physically showing or passing around their phone to friends or uploading them to online networks. These practices effectively create hybridized modes of communication that cut across mediated and co-present or face-to-face contexts. Yet perhaps of most significance in relation to quotidian sociality and practices of image-making-and-sharing is the exponential rise of the selfie as a mode of mediated and embodied communication. As the mobile device became an ever-ready conduit to social media, the intimate relationship between the mobile screen and the face intensified, congealing in the phenomenon of the selfie as people engage in the performative exhibition and curation of their own lives. Despite the criticism often levelled at the narcissistic tendencies of the selfie-taker and social media more generally, the practices of self-portraiture – the sharing of our faces *en masse* – has also fostered a collective sense of intimacy or imagined "being-together." As Zilio comments:

> the diffusion of anonymous self-portraits across networks reinforces a feeling of participation in the world, by symbolizing individual lived experience in the style of the collective. This construction in the present of

> a real heightened by moments that are strictly speaking unimportant – selfies in elevators, rear-view mirrors, or public toilets – mobilizes an imaginary of the everyday that is familiar and shared by all ... the inauguration of gestures and attitudes that are universally and unanimously shared. (Zilio 2020: 117–18)

Kenaan observes that selfies almost always include an identifiable face (or otherwise they are "not really" selfies, or typically tagged as #selfiefails), and are by definition shared; they are a mode of making one's face (or self) "present in the public domain" (Kenaan 2018: 114). Moreover, selfies are performative of a specific facial ontology, as we are not able to see our own faces without some form of material or technological affordance: "our face escapes us. It is not given directly to sight, and its integration into our field of vision requires the mediation of an image. This mediation is [now] deeply rooted in our daily visual routines" (Kenaan 2018: 122). Selfies are mobile snapshots of a moment, a facial expression, a place, an action or mood, frozen in time, intended to be shared and circulated. Once taken they are uploaded to social media platforms, awaiting likes and comments from followers and other users, in a digital economy where "sharing" has itself become a form of currency. Each time a selfie is posted, the image – usually featuring the face – becomes an exchange, part of the platform's data, used to train algorithms and inform marketing strategies. Such acts of sharing are the predominant activity of everyday users in the larger system of digital labor, as the selfie continues to circulate as one tiny element in digital archives of billions of faces. But for the user, the selfie remains personal, representing a small act of creativity, identity and self-expression, even as it contributes to the social media cycle of commodification.

Phenomenologically, the selfie demands a coordinated positioning of the body – arm outstretched with an invisible "hand–camera assemblage" turned towards the face, often requiring a certain amount of practice and "limbic and manual dexterity" (Frosh 2015: 1612, 1614). The selfie also invokes a particular relationship between the body (and more specifically the face as metonym of the self), the social network and the mobile device. Frosh calls this body–technology relation a form of "kinesthetic sociability" (2015: 1608), as taking and sharing a selfie connects bodies, their hybrid mobility through actual and online spaces, and the "micro-bodily hand and eye movements" as we daily engage in the acts of sharing, linking, tagging and swiping that sustain the prolific selfie economy.

Conclusion

Throughout this chapter we have considered the various body metaphors attributable to screens, and the problematic assumption that the window and frame are perceptually homologous to contemporary media screens, particularly in relation to mobile media. We explored our "facial" engagement with screen interfaces and suggested that there is a certain material and physical kinship between faces, windows, frames and screens in terms of the way we "turn" to them with varying degrees of attention and distraction. The mobile interface also has a special affinity with the face in terms of the prolific rise of the selfie, and its role in our everyday practices of mediated intimacy.

The "telic inclination" of the screen is not uniform, linear or continuous, or necessarily determined by the perspectival trope and its demands for a fixed face-to-face relationship. With a more nuanced phenomenological analysis of the micro-practices surrounding our experience of contemporary screens, we can more effectively interpret the way mobile devices modify our communicative and playful practices, remediate our experience of media content and insinuate themselves into our ways of being-in-the-world. The mobile media device, to a degree at least as significant as the cinematic, televisual and computer screen, presents a significant shift in the relational ontology of body and technology. This relation is perhaps more intimate, ever-present and affective than any we have thus far experienced. What we need, then, are ways of thinking through new body–screen metaphors that more effectively capture the distracted, discontinuous, motile, peripatetic and tangible nature of mobile media engagement.

As theorists such as Ihde and Friedberg have pointed out, ways of encountering the world, both mediated and unmediated, entail conventions of sense perception and collective corporeal habits that are not innate or given, but culturally, materially and bodily specific. Each new interfacial modality stipulates its own gathering of *soma* and *technique*, its own embodied routines. In this light, we have suggested that our contemporary media experience unhinges preceding face– and body–screen couplings. That is, the particular body–media configurations of screen experience across televisual, computer and mobile interfaces, when critically examined in terms of their medium-specific effects, can offer some insight into how such effects work to confound and reshape historically sedimented face-to-interface conventions.

2
Eyes

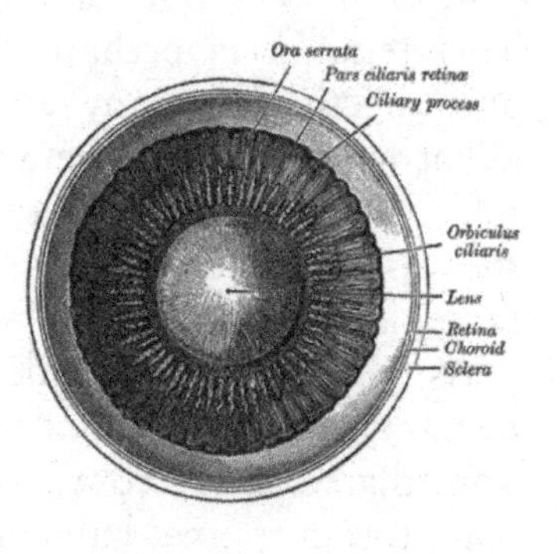

The primacy of the eye ... as the dominant sense organ of the twentieth century is a partial effect of a technical revolution that put an enormous apparatus to the service of vision. The rise of the eye is rooted in the fact that all of its aspects (creation, transmission, reception) were supported by analog and digital machines. The triumph of the visual in the twentieth century is the triumph of techno-vision. (Weibel 1996: 339)

The eyes are one of our most complex organs, consisting of over twenty parts – the primary ones being the cornea, iris, pupil, aqueous and vitreous humors, retina, optic nerve, crystalline lens, anterior and posterior chambers, sclera (or white of the eye) and blood vessels. They are particularly vulnerable to injury and so are protected by the orbit or socket which is comprised of skull bones that form a four-sided pyramid pointed at the rear of the head, and the eyelids which distribute lacrimal secretions, sweep debris away and retain necessary moisture (Perkins 2023). Light enters the eye through the cornea and pupil, is focused on the near or distant field by the lens's surrounding ciliary muscles, and then translated by the retinal rods (for light and dark) and cones (for color) into an electrical signal that travels from the optic nerve to the brain. Human and mammal eyes are often "metaphorized" as "camera-type eyes," as the cornea and lens are considered similar in function to a camera lens, and the retina analogous to film (Helmenstine 2019). Like a camera, the initial image on the retina is inverted, and then flipped right way up by the brain. Such a technological trope for human body parts follows a long tradition of conceptualizing embodiment in terms of our tools: the human body has often been conceived in terms of the dominant technologies of the time, including as a clockwork mechanism, as a motor or machine, or as electronic circuitry and data.

The eyes and vision are perhaps the richest source of body metaphors. The eyes are often referred to as the "windows to the soul" and are the facial feature most closely linked to – and revealing of – the self or personhood. Along with touch, our most intimate interaction with other humans is through eye contact. In both historical and contemporary thought, science and everyday life, vision is also considered the most "elevated" sense, and the closest to knowledge, intellection and truth. As Romanyshyn insightfully observed over thirty years ago, the primacy of vision has become "a habit of the mind"; so many of our technologies and media interfaces "emphasise this feature of visibility ... we might venture to say that our sense of reality has nearly become identical with our ability to render something visible" (1989: 184). The predominance in Western thought of the visual metaphor, combined with an expansive range of visually oriented social and technocultural practices in both image- and screen-based entertainment and scientific discourse, is indicative of the close relationship between vision, visuality and human understanding. Our culture is so deeply sedimented in this visualist tradition that the emphasis on the visual appears not as a bias, but rather as completely natural and given.

In this chapter we trace the history of ocularcentrism – the primacy of vision and the "triumph of techno-vision," as Weibel (1996) so aptly puts it – in the context of media screen practices, and consider the role of sight in the eyes–hands–screen circuit that typifies our use of mobile media. We provide a range of examples from location-based games and our experience of Google Glass and Spectacles by Snapchat, in terms of how mobile eyewear complicates the cultural contexts of the "look." Following our central argument, we show how the body–technology coupling of the eyes and mobile technologies is ontologically "primary," comprising a relationality that offers unique affordances in terms of determining and transforming our experience of the world.

The Primacy of Vision

> How is it, we might ask, that vision came to seem so apt a model for knowledge? [...] If we cease to accept

> the visual metaphor as necessarily natural or intrinsic to the meaning of knowledge, then it is essential to [...] ask what particular relation between us as knowers and the nature to be known is implied by such a metaphor. (Keller and Grontkowski 1983: 208)

Western notions of knowledge and knowing situate vision as the most direct access to truth – that which can be seen is "real," described by many familiar tropes such as "seeing is believing" or "I see what you mean." We might say that both vision and light sit at the apex of sensory perception, and the privileged place of vision in Western thought is based on the idea that vision is a universal and natural attribute, providing humans with unmediated and candid access to an objectively present world. Knowledge itself is described as illuminating or enlightening, and the process of coming-to-know is often synonymous with using one's eyes. Vision is understood as a primarily neutral and transparent mechanism which accesses inherently perceivable objects. Moreover, it is the supposedly impartial means through which we "grasp" knowledge about the world; no other sense is as trustworthy, objective and reliable. This deep connection between vision, light and knowledge is so embedded that we often forget that it is a very powerful metaphor with particular perceptual biases.

In their well-known feminist critique of Western knowledge, *The Mind's Eye*, Keller and Grontkowski (1983) suggest that knowledge based on vision is premised on the idea of a disembodied intellect, and that other modes of "proximal" perception – such as taste, touch, smell and hearing – could not have claimed the purity of "objectivity" as they are sullied by intimacy and individual differences. Any such change to our sensorium is also unlikely, as cultural variations to sensory organization have significant historical density behind them. As Marks (2000) suggests, a new configuration of the senses would be meaningless without the sensory and collective memory to accompany it. The slowness of changes to our cultural and collective sensoria, however, does not mean that flexibility in our corporeal schema or the spectrum of sensory possibility is not possible. Indeed, as we suggest below and throughout the book, non-visualist possibilities can and do exist in mobile media practices, and reside quite comfortably (though paradoxically) alongside the ocularcentric narrative.

Nevertheless, in both scientific knowledge and philosophical enquiry, the eyes have continued to serve as the organic prototype of knowing, such that *observability* is conflated with *knowability*. This primacy of vision is also implicit in quotidian life, where the status of vision as the best and most valuable of the senses is played out in a vast number of visual metaphors and directives ("see what I mean?," "is that clear?, "you need to focus!," "look me in the eye"). From early Greek philosophy onwards, questions concerning the acquisition and boundaries of knowledge have frequently been preoccupied with visual evidence, and thus the existence of things in the world often becomes similarly confined to that which *can* be seen. The historical weight of this link in Western thought can be traced to Plato's view that, "if light is a thing of value, the sense of sight and the power of being visible are linked together by a very precious bond, such as unites no other sense and its object" (cited in Harding and Hintikka 1983: 210).

Yet many scholars have demonstrated that there is no universal, essential vision or way of seeing but many culturally, historically and *medium*-specific visualisms. As noted in Chapter 1, even the seeming intuitiveness of perspectival vision was "naturalized" by the mechanism of Alberti's grid. Other apparatuses have introduced equally specific modalities of seeing: the development of the microscope, for example, brought discrete minutiae into visual experience; the emergence of impressionist and surrealist art also challenged the universality of a three-dimensional and objective world-landscape; and, possibly of the greatest significance within everyday life, photography's ability to frame the momentary, the fragmentary and the arbitrary disturbs the very notion of a uniform, continuous and holistic perspective. Indeed, the inherent manipulability of the digital image calls into question the documentary "truth status" of photography, not to mention the current debates around deep fake technology. Yet despite these challenges, the pervasive supremacy of ocularcentrism that links vision to truth and knowledge has not to date been significantly destabilized in either scientific or common-sense understandings of what constitutes objective reality. Or, perhaps more accurately, we are able simultaneously to maintain a belief in this metaphor while paradoxically acknowledging that the vision–truth connection is frequently "broken" by multistable cultural contexts, individual differences and the capacity of new technologies to undermine the veracity of the image.

Regardless of this contradiction, the "hegemony" of vision is prevalent in the technological development of visualizing and imaging devices, where the primary aim is to overcome all obstacles to vision, be they barriers of size, opacity or inaccessibility. In Ihde's (1991) analysis, this trajectory has been carried by our belief in *instrumental realism*, where captured visual images are taken as representative of the "real" (though, as noted, we enact the paradox of disbelief in digital "truth" at the same time). In other words, the bias towards ocularcentrism is endemic in our use of technology; most of our technological devices in both scientific contexts and everyday life – including the lens, telescope, microscope, MRI and CT scans, television, cinema, photography, mobile screen and augmented or virtual reality devices – either visualize phenomena or translate non-visual information into visual form. They partially determine what *can* be seen and *how* it is seen. In both popular culture and scientific discourses, perception is now so deeply entwined with imaging technologies that it's impossible to conceive of either scientific or common knowledge outside of its embodiment in visual instruments designed to extend and augment the perceptual reach of the eye.

Screens, Mediated Vision and the Mobile Camera

> When the window of linear perspective vision has become the primary cultural metaphor, a habit of the mind, the world has become primarily a matter for the eye alone. It has become primarily a visible matter, well on the way towards becoming a bit of observable, measurable, analyzable data, readable as a computer print-out, for example, or perhaps as a blip on a radar screen. Indeed, so many of our technological instruments emphasise this feature of visibility – microscope, telescope, camera, television – we might venture to say that our sense of reality has nearly become identical with our ability to render something visible. (Romanyshyn 1989: 184)

The technical enhancement of the eye is often about revealing what is not see-able, of making accessible that which is not immediately visible to oneself or others (but through mediation can be made available), enacting the displacement of vision from the human eye to the materialist veracity of the visual *mechanism*. Each mode of technological revealing demands or affords a particular type of embodiment, or specific modalities of the body–technology relation, in Ihde's terms. For example, consider the case of telescopic sight. Firstly, as the term "focus" implies, when something is brought *into* focus, one focal length is chosen at the expense of others, such that other objects are backgrounded or extracted from vision; focal range brackets out much of the environment in order to see a particular object/scene with visual acuity. Secondly, telescopic sight flattens out depth, such that what is seen appears to have a two-dimensional or depthless quality. Thirdly, the lens frames the object, which decontextualizes its environmental "situatedness," while reflected light on the lens can cause back glare, and flecks, stains and cracks on the glass may generate visual anomalies or "artefacts." Finally, the telescope changes the eye–body relationship; while a distant object may seem close, it cannot be touched, heard or smelt. We might say that a telescopic object is attained by a particular kind of medium-specific tele-body; a long-sighted, one-eyed body without auditory or olfactory function.

As this example reveals, in *transforming* the "what" and "how" of seeing, visualizing technologies show us that visual perception is not only mediated, but highly variable or multistable. In some ways we might think of them as active "partners" in the perceptual process, to borrow from Haraway's (1988, 1991) notion of technological agency. Yet this multistability is at the same time culturally and somatically specific, affected by collective and individual differences. Oliver Sacks (1985, 2010) has famously documented the "plasticity" of visual perception through empathetic descriptions of a range of neurological conditions including color blindness, visual agnosia (inability to recognize faces and objects), and loss of sight in one eye resulting in a "flattening" of depth perception, while people on the autism spectrum often experience increased sensitivity to motion and heightened pattern recognition. In the modern world, we have collectively "learned" linear perspective over centuries of Cartesian and Euclidean representations of three-dimensional and geometric space (Heelan 1983).

The multistable and disparate nature of contemporary vision is the partial effect of the many screens encountered in the everyday lifeworld – televisual, cinematic, mobile, closed-circuit, large and small – each with their own technical, environmental and interfacial specificities. Ihde (1993) calls this mode of seeing "plurivision." Upon waking, a person in Melbourne (Australia) may immediately check their phone for messages and notifications, watch the morning news on the large flatscreen in the living room, then travel to work or study through an urban space in which screens are now ubiquitous – from advertising billboards and ATM touchscreens to the projection of artwork onto buildings in Federation Square – effectively turning the built environment into a screenic surface. The computer screen becomes the primary conduit of information and communication during the average weekday, while back at home in the evening the smart speaker screen responds to voice commands and the tablet becomes an interface for online play. Throughout the day, the Apple Watch or Fitbit screen provides ongoing updates on one's personal movement, tracking steps and biometric data.

Today, screens and cameras are visual tools that have become virtually inseparable from what we would discern as our "own" perceptual and sensorial boundaries, and have given primacy to a *particular* kind of vision which can be said to translate the world *as picture* – in which we as subjects take up the privileged position of organizing and sharing visually biased interpretations of the world. Marks (2000) documents the neurophysiology of sense memory (via Varela and Rosch), and its embeddedness in social memory, in an analysis which can account for the way in which sense-augmenting technologies are gradually integrated into our sensorium. Our experience of mobile media entails an implicit awareness of the way it mutates and reconstitutes our visual situation, while our use of the mobile camera is sedimented (if we are old enough) in our familiar and habitual experience of the standalone camera before it, and our prolific viewing of screen media more generally. Consider, for example, the number of "synthetic" ways of seeing the cultural interface of the camera has implanted in our vision: the wide-angle shot, allowing for "maximal digestion" of the scene; the zoom, a constructed vision organizing its space "through the force of intention alone, unmired by distance, time, and its inherence in things" (Singer 1990: 60); the close-up, where we can witness something at an intimate proximity that would otherwise be disrupted by our actual presence; the split-screen and jump-cut, which establish semantic and temporal cohesion between elements that are otherwise physically distant

(Singer 1990: 61). All of these configure non-normative body–space relations, but nevertheless ones we have "learned" collectively by way of film and television watching and our long history of camera use.

In her description of what she terms the "contagion effect" in our experience of visual media, Singer comments on the "pleasure of sociality, the moment in which one's pleasure in seeing is enhanced by its reproduction and affective reverberations in others" (1990: 55). Although less structured and confined than cinema, the mobile media experience is also "contagious" in the sense that it is implicated in a dispersed collective embodiment – an *intercorporeality* in phenomenological terms – which is the foundation of vision per se, and structures the specific visual situation of telepresent vision. Our sensory expectation is thus directed towards the mobile interface in what becomes a collective and habitual manner, and in the context of mobile phone photography we encounter it equipped with a sensorial-social "memory" of screens and the cultural interface of the camera: we anticipate the device's synthetic ways of seeing, we orientate our bodies in such a way that visual acuity is best served, and we know that it offers a compound vision, a sometimes-linear, sometimes-jumbled stream of images with sound, but without smell or textural dimensions other than what our own sensorium can project. That is, each technological embodiment of vision gathers the human body into (yet another) somatic involvement, be it one which prioritizes hand–eye coordination or one which fixes the body's position in front of a screen, thus also reconfiguring or privileging a particular spatial ontology, a mode of spatial arrangement and containment in the world.

Mobile smartphones have become the contemporary camera *sui generis*, and are now a mundane yet significant part of visual and participatory media cultures; frequently used to make "evidentiary" recordings of everyday happenings – criminal, comedic, intimate and affective events shared on social media – the mobile device becomes a form of prosthetic vision, both mirror and container of contemporary life. The mobile phone provides us with a complex aggregate vision, a continuous slippage and merging between genres of visual meaning that requires a literacy of imaging techniques, such as the freeze-frame, the close-up, slow-motion, time-lapse, zoom and panoramic view, and an awareness of the variable capacities and functionalities of digital imaging, including "view outside frame," "high dynamic range," "burst mode" and filters. The mobile camera phone has irretrievably and profoundly transformed the contemporary visual terrain of perception and experience. Such modalities of seeing are also platform specific, as exemplified by TikTok dance-challenge videos, a participatory form of embodied sociality achieved through stylized "micro-performances" that have a collectively understood vocabulary or choreography (Griffiths 2023). As Griffiths notes, the effect is one of "kinaesthetic empathy" as the audio-visual performance is both seen and internally simulated or "felt" in the bodies of other users, evoking a sense of corporeal connection and belonging (2023: 71).

Haptic Vision

Against the traditions of ocularcentrism, the association of vision with touch and the influence of the haptic sense on our visual perception has a long history. For instance, according to Gandelman, it was Berkeley who posited that "there is no vision in the performative meaning of the term – that is, in the sense of seeing as a potentiality of acting over the objects that surround us – without a transfer of the sense of touch to the properties of the eye" (Gandelman 1991: 6). The Italian futurist Marinetti put the matter more directly, declaring in a 1924 manifesto on "tactilism" that "a visual sense is born in the fingertips" (1991: 120). Similarly, Irigaray argues that without a sense of touch, seeing would not be possible (Vasseleu 1998: 12). Indeed, we learn three-dimensional vision and perspective always *after* our motile, tactile, haptic and aural engagement with the world: "tactile sense is correlated with our allegedly two-dimensional retinal image, and this learned cueing produces three-dimensional perception" (Heelan 1983: 189).

Such an understanding of haptic vision can shed light on our bodily orientations to mobile media, especially in terms of the more recent ubiquity of touchscreens. Haptic engagement works because vision completes the illusion (for example – in the case of mimetic mobile games such as *Angry Birds* – of gravity and movement behaving according to expectation). It can also contribute to a fuller comprehension of what Cooley has characterized as an increasingly "material experience of vision" by mobile media users, where "hands, eyes, screen, and surroundings interact and blend in syncopated fashion" (2004: 145). More literally, haptic vision is also a feature of vibration functionality and mobile accessibility apps designed for people with visual impairments, allowing users to "see" with their hands (for more detailed discussion see Chapter 4).

In transforming the physical environment into a playground, location-based games force mobile gamers to divide their attention between two things: the mobile interface and the information the device offers, and the actual physical setting in which the gamer is moving. Involved here is a canny and subtle form of environmental knowing in which one is attuned to the specific requirements of mobile gameplay while also retaining a crucial peripheral awareness of the spatial surroundings (Hjorth and Richardson 2020). Performing this double requirement involves a particularly adroit oscillation between stickiness and distraction. In this respect, the "ways of looking" specific to cinema, television and console games – characterized by the gaze, the glance and the glaze respectively (Chesher 2004) – does not translate accurately onto mobile phone conventions of screen engagement, which demand adaptable and dynamic levels of sensorial detachment, "stickiness" or immersion. The mobile phone device crosses over each of these ways of looking if only because we can – and do – experience many forms of content on our phones. Or, at least, we might identify a broad spectrum of corporeal "attachment" and "distraction" across a range of practices – from games and social media, to map navigation and interpersonal text- and video-based communication – based upon levels of immersion, engagement and distraction. In sensory terms, recalling the opening vignette of the book's Introduction describing *Pokémon GO* play, location-aware mobile gaming encourages tactile vision; that is, a material and dynamic seeing that involves eyes, hands, feet and the mobile screen device. It undoes the primacy of vision typified by Alberti's grid and supplements the eye-ear predominance of audio-visual screens with whole-of-body or macro-body movement.

In the following section on eyewear, we consider what is significant about augmented reality as a form of perceptual mediation in terms of the way it radically alters the visual relationship we traditionally have with screens, quite literally transforming what becomes visible and the process by which it becomes seen, requiring new "techniques of the body," to use Marcel Mauss's (1973) famous phrase.

Eyewear: The Case of Google Glass and Snapchat Spectacles

The development and ultimate demise of the first Google Glass prototype provides an instructive case study of the relation between visual perception and innovation in mobile technology, not least because it represents an early introduction of augmented reality glasses to the public, and highlights how the usability and affordances of new mobile media ultimately fail when they are at odds with the collective habitudes and expectations of mediated vision.

Google Glass is a wearable computing device in the form of a pair of head-mounted clear glasses with a heads-up display attached to the top corner of the right lens. Google's introduction of the first Glass was cautious: the company pursued a "soft launch" in 2013, with prototypes released to small cohorts of "Glass testers" comprised of technology early-adopters, engineers and journalists, followed by more widespread availability the following year. A newer prototype developed in 2022 – also now discontinued – saw a more strategic and staged release, with marketing focused on speech recognition rather than the gimmicky applications mired in privacy concerns that plagued the first version (Robson 2022).

In certain respects, Google Glass and augmented reality glasses in general are disconcerting for the very reason that their familiar form belies their less familiar function. In design terms, the first Google Glass prototype constituted a subtle variation on a very personal yet now accepted and habituated technology: glasses. Glasses, or spectacles, are widely understood as a vitally important technology ("one of the most useful appliances known" [Rubin 1986: 321]), yet are so familiar as to be largely taken for granted. A tension between familiarity and technological fascination is evident even from the earliest pictorial representations of glasses (Daxecker 1997; Rubin 1986: 322). Indeed, this tension structures the entire history of the development of glasses and follows a pattern of continual technological innovation – such as Benjamin Franklin's invention of bifocals in 1784, followed by subsequent adjustment and wider societal adoption (Cashell 1971; Rubin 1986). At a more intimate scale, this specific body–tool coupling is far from settled; the bodily adjustments (or lack of them) that follow with newly acquired glasses are well documented. For example, Lord et al. (2002) report on the risk of injury from falls, and the consequent macro-perceptual defamiliarization of one's environment and diminished sure-footedness, that result from the loss of depth perception and edge-contrast sensitivity among older wearers of multifocal lenses. In the case of Google Glass, a brief internet search reveals how, for individual wearers, the intercorporeal experience was one of continuous perceptual negotiation and adaptation *in situ*.

What was significant about Google Glass as a mediating device was its streamlined interface which all but effaced the frontal face-to-face or eye-to-screen ontology of other familiar body–screen relations, combining augmented reality and embodied emplacement in ways that sought to quite literally "transform the landscapes around us into information interfaces" (Farman 2012: 43). The display of data in front of the eye, rendering the screen and the frame invisible, *and* the capacity to capture data in concert with one's head movements, made it a perceptually different experience from other head-mounted or handheld augmented reality tools. In this way, Glass hardware and software (Glassware) required new "techniques of the body," including gestural actions and reactions such as single and double finger taps, eye winking and voice activation, as well as a broader suite of micro-muscular and near–far visual adjustments on the part of the Glass wearer-user.

In an article in *The Atlantic*, de Zengotita (2014) took issue with the body–technology defamiliarization that Google Glass demanded of wearers. He argued that "the containment of the frame, the *placement* of the screen on a device," is deeply ingrained in the way that "our bodies are oriented" towards media technologies and our ability to "face or turn way from other things" that surround us. Google Glass's removal of the "frame" that contains our screenic interactions, he suggested, severely disrupts this schema. While significant, de Zengotita's critique perhaps overstates both the "contained" nature of mobile devices and the visual ontology of the interface. As we have argued elsewhere (Richardson and Wilken 2012; Richardson 2010), and noted in the previous chapter, the mobile phone screen and handset already disrupt the frame-based and frontal ontology of other screen-based interfaces, changing the way we "attend" to, capture and memorialize people and places.

Viewed in this way, the first Google Glass was the latest in a line of ocularcentric technological developments requiring varying degrees of bodily adjustment on the part of those using them. Having said that, the Glass case is striking in a contemporary context for the specific body–tool reconfigurations it brings about, and for the larger disruptions to accompanying social norms and expectations it provokes. For instance, the initial coverage of Glass, within both the trade and popular press and academic scholarship, was dominated by concerns over the privacy implications of users being able to take photos, videos and livestream audio-visual footage through the Glass interface. There were numerous cases of Google Glass being banned from venues, and isolated incidents of public disturbances between users and passers-by (Meese 2014).

These anxieties and tensions emerge from the way that, in addition to the gestural and other bodily adjustments, Google Glass reconfigured our experience of vision and ways of seeing. The device presented for the wearer and those watching them the same phenomenological question that Merleau-Ponty once posed: "Now what actually is fixing one's gaze?" (2004 [1962]: 263). For the wearer, the projection area of the right lens separates the region of focus from the surrounds and thus interrupts the larger field of the spectacle. For those watching the wearer, the intimate scale and facial positioning of the device, and the proximity of the projected data to the eye of the wearer, make it difficult to ascertain what the wearer is focusing their attention on, or, indeed, what one should take as the *intentionality* of the look. One of the specific anxieties about the Winky app for Glass, which enabled the wearer to record what they see simply by winking, is that it opened up the possibility of gestural confusion and the misreading of facial cues and orientation, while also generating uncertainty – not simply about what the wearer is looking at (which is already the case with sunglasses that have a dark or reflective surface), but as to whether the look is transient (of the moment) or indicates the more permanent capture and potential on-sharing of a photographic image or video. What Google Glass prompted, in short, was a reconfiguration of the "total structuralization" (Merleau-Ponty 2004 [1962]: 263) of the field of vision for both the wearer and their near-dwellers, disrupting our collective patterns of media use and perceptual proclivities.

Snapchat Spectacles, launched in 2016 by Snap, have had a somewhat different evolution, as from the start they were developed for exclusive use with the Snapchat app for mobile devices, which was already a "mundane" and familiar means of creating and sharing personal everyday experiences for over 350 million daily users. The Spectacles are much like sunglasses – thus more "wearable" than Google Glass – with integrated camera and microphone that records video of up to thirty seconds by tapping a button on the side of the frame, connecting to the user's smartphone via Bluetooth and enabling automatic syncing with a Snapchat account and subsequent sharing with friends. The camera lens has a wider field of view than a phone camera, and captures video in a circular format, which is intended to more closely "approximate" human vision (Newton 2016). Interestingly, concerns over privacy were partially addressed by the integration of a light that appears on the glasses when recording, an addition that was described by reviewers as a "responsible design choice" (O'Kane 2016). Talking about his first experience of the glasses, Snap CEO Evan Spiegel describes a more-than-visual or enhanced sense of "felt" presence, claiming it was like seeing "my own memory, through my own eyes ... It's one thing to see images of an experience you had, but it's another thing to have an experience of the experience. It was the closest I'd ever come to feeling like I was there again" (cited in Newton 2016). The Spectacles 3, released in 2019, added a second camera that affords depth perception and the addition of 3D augmented reality filters. Spiegel (cited in Newton 2019) acknowledges that widespread uptake of AR glasses is years away, yet believes they will follow a similar trajectory to the evolution of cameras, which were initially "event-based" – that is, used to record special occasions – but are now thoroughly embedded in quotidian media-sharing practices and the seamless memorialization of everyday life.

The development of Google Glass and Snapchat Spectacles both follows and departs from "sedimented" and evolving trajectories of mediation, body-media tropes and collective habits. These include, for example: the familiarity of glass-wearing we bring to the experience, which is itself part of the larger histories of both visual and wearable technologies; the perceived affordances of glass – an architectural substance that elicits its own tensions between transparent openness and observational control (McQuire 2003) – against the communicative and mediatic enablements of AR glasses as patented media interfaces; cultural imaginings of virtual and augmented-reality cyborg futures unevenly realized in wearable media technologies; the perceptual and metaphoric legacies of screen-based interfaces and the more recent transformative effects of networked and mobile media; social mores relating to image-capture and sharing, privacy and personal communication; and finally, the long, long histories of navigational, peripatetic and place-making practices that diverge along new paths formed in part by our uptake of location-based and mobile media, and so on (this is by no means a comprehensive list).

These situated, intercorporeal, evolving and dynamic ontologies clearly show the irreducible relation between specificity and process out of which AR glasses have emerged. As illustrated by Moores (2012, 2013), Couldry (2012), van den Boomen (2014) and others, a non-media-centric approach allows us to critically interpret the ways our media metaphors inform and are informed by a myriad of extra-medial experiences in the everyday lifeworld. van den Boomen writes: "Metaphors such as virtual community, Web 2.0, Facebook friend, following, phishing, and liking all tap from other resources. They translate and transform conventions, acts, habits, and desires into digital-material entities that subsequently become mediatized" (2014: 191). In our corporeal apprehension of Google Glass there is clearly a complex assortment of habits, tropes, conventions and affects, some of which are related to media and screen engagement, while many others originate from a broad spectrum of body–world relations. As always, these relations must be considered in terms of both their situatedness (i.e. as instances of body–tool couplings) and their more enduring effects on our collective habitudes and practices.

The first prototype of Google Glass was discontinued two years after launching, deemed a failure due to tech glitches, unappealing aesthetics, battery issues, health concerns (such as constant exposure to carcinogenic radiation), privacy and piracy risks, and not least because the "why" or value of the device was not clearly presented to consumers – even the developers disagreed on how often and when it should be used (Leonard 2022; Yoon 2018). At the time of writing, the later Google Glass 2 Enterprise edition has also been discontinued amid mass lay-offs at Google in response to recession fears. The first iteration of Snapchat Spectacles also had limited market success – in part due to their restricted use with the Snapchat app – with Snap writing off $40 million worth of unsold inventory in 2017, though later generations were developed that enabled exporting and sharing across platforms (such as YouTube VR), and they continue to be on-sold through online marketplaces. Snap's latest Spectacles allow users to view AR objects through the lenses – thus including functionality integrated into Google Glass – though these are not for public sale but instead distributed to creators and developers "looking to push the limits of immersive AR experience" (spectacles.com/uk). Despite their rather checkered emergence, innovation and development in augmented reality glasses continue to flourish, with both Apple and Meta expected to launch glassOS and Meta Glasses in the next few years.

Such interfaces undoubtedly demand new modes of experiencing and negotiating our being-in-the-world that are still very much in flux. Their habituated use will require subtle yet significant reformulations of and adjustments to the established body–technology couplings that pair people with glasses, faces and eyes with screens, and bodies with media, together with the complex social implications of these reconfigured pairings. Five or ten years from now, we may well have adopted the "habit" of augmented reality glass-wearing in everyday life, recording every moment of our waking lives in point-of-view 3D video embedded with context-appropriate filters and effects.

Conclusion

Our experience and perception of time, mobility and distance, and the material parameters and limits of experience, have mutated and adapted to "tele-technologies" such as the telephone, the television and network media – technologies that enable perception at-a-distance. Media theorists such as McLuhan (1964), Innis (1984) and Carey (1969) have long argued that such technologies have had gestalt-like effects on cultural practices, the physical formation of societies, the manner in which we acquire knowledge, and on the ways in which we understand the body and its limits. The mobile media interface, to a degree at least as significant as other screens, presents a shift in the relation between the body and technology, and, more specifically in the context of this chapter, in the complex mediated relation between vision and the screen. Mobile interfaces are exceedingly intimate and ever-present in many aspects of our emotional lives and everyday mobilities, offering a "personalized" and customized visual slice of the world. As noted in the previous chapter, in a very fundamental way they change what we pay attention to, and the modes and duration of that attentiveness. The proliferation of mobile media and mobile phones has created a culture of constant connectedness; we are increasingly reliant on them not only to communicate but also to navigate and experience urban environments. They have profoundly transformed visual perception, providing us with immediate access to a wealth of visual information, ranging from maps and playful apps to shared photographs and videos, affording us the capacity to experience the world in local, "glocal" and a myriad of digital-material ways.

If Western knowledge has instilled a perceptual hierarchy where tactility, haptic proximity, motility and sound have been rendered subordinate and secondary to the act of seeing and the criteria of visibility, it might seem that modern visualizing technologies, in their work as perceptual devices, further this elite visual bias, the consequence of which has been a marginalization of the haptic, the tactile and the kinesthetic. Yet mobile screens as locational and haptic interfaces directly challenge this notion. That said, from a postphenomenological perspective even so-called "unmediated" vision does not provide us with privileged access to the truth of things, and technologies of seeing and imaging are not distal and non-intervening modes of perception removed from the corporeal entanglements of the body. Even within the everyday practice of watching screens, mobile or otherwise, we are not transported by vision away from the body. Rather, our somatic and cultural memories, our sense of collective embodiment, our embodied navigation of mediated space and of the body–technology interface all work to make the experience quite full-bodied. Mobile media, in particular, modify the eyes' relation to the arm and hand, such that haptic perception is also altered. That is, the focus-function of such devices – and indeed of all forms of remote control or telepresence – demands a kinesthetic awareness of the tactile screen and its effects on our ways of seeing. Before turning to a more detailed consideration of touch and haptics in Chapter 4, in the next chapter we explore the sensory domain of the ear, and the many ways mobile interfaces have remade our aural experience of the world.

3
Ears

[T]he order of listening which any telephonic communication inaugurates invites the Other to collect his whole body in his voice and announces that I am collecting all of myself in my ear. (*Barthes 1991: 252*)

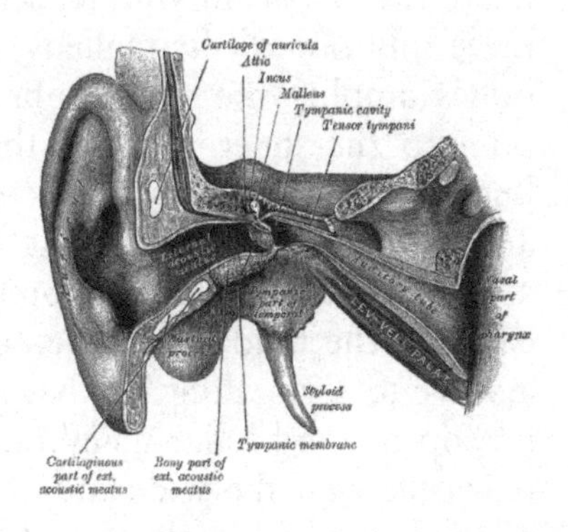

Somewhere in the middle distance a dog barks. How is it that we humans not only register that this sound is that of a dog barking but are also able to approximate its direction and its distance from us? Sound travels through the air as pressure waves. As these pressure (sound) waves reach us, our outer ears – our pinnae or auricula, the skin-covered cartilage shaped into twists and folds – serve as funnels "that collect, amplify and direct sound waves to the ear canal" (Scientific American 2015). The human pinna (ear) also helps us to determine sound direction. Sounds from the front or side are enhanced, while those from behind are reduced, producing small differences in volume. These modulations in volume, together with minute distinctions in the time it takes sounds to arrive, help us to determine the location and distance of a sound's source (Scientific American 2015).

Once these pressure (sound) waves enter our ear, they travel through the outer ear canal to the middle ear and the ear drum. The incoming sound waves make the ear drum vibrate, sending these vibrations to three tiny bones (the malleus, incus and stapes). These bones amplify the sound vibrations, forwarding them on into the inner ear and the snail-shaped structure known as the cochlea. The cochlea is filled with fluid and is divided along its length by a partition known as the basilar membrane. Sound waves detected by the ear lead the fluid inside the cochlea to ripple, causing a wave to form along the basilar membrane. Hair cells on top of the basilar membrane ride this wave. Those hair cells near the wide end of the cochlea detect high-pitched sounds, such as a baby crying, while those towards the center detect lower-pitched sounds, such as a dog barking. When the hair cells rise up and down, "microscopic hair-like projections (known as stereocilia) that perch on top of the hair cells bump against an overlying structure and bend. Bending causes pore-like channels, which are at the tips of the stereocilia, to open up. When that happens, chemicals rush into the cells, creating an electrical signal" (NIDCD 2015). These electrical signals are carried by the auditory nerve to the brain where they are converted into a sound that we recognize and comprehend – in our case, that of a dog barking in the middle distance somewhere to our right.

A capacity to hear is something that most humans have in common, although this capacity (as we explore in greater detail below) is certainly not experienced in the same way by each of us – it differs between individuals and across cultures (see, for example, Ochoa Gautier 2014). Clearly those who are Deaf or Hard of Hearing experience sound, hearing and listening very differently from those with unimpeded hearing capacity. For the former, some may have extremely restricted hearing, while others may hear specific frequencies or sounds within a certain volume range. For those who are able to hear, the capacity to discern particular sounds at an individual level can be shaped by individual knowledge and skill, among other factors. For example, an experienced bird enthusiast may be able to identify and distinguish between different and related species by listening to a bird's call, whereas a novice may or may not even register the presence of a bird, let alone differentiate its call from others that are similar or nearby. Our capacity to interpret sounds both individually and collectively can also be context specific. In the game of association football (soccer), for example, it is common practice for defenders to be aurally guided in their actions by their goalkeeper, who is stationed behind them; the call "right shoulder," for instance, alerts a defender to the presence of an unsighted opponent approaching them from behind and to their right side. Our capacity to interpret sounds at an individual and collective level is also culturally shaped. Visitors to Ho Chi Minh City, for example, can be overwhelmed by traffic noise and the seemingly indiscriminate and constant sounds of honking horns. Those with local knowledge, however, are more likely to be aware of the subtle differentiations between forms and combinations of honking; these comprise a complex aural system of notifications and signals of intention to other traffic users (for a playful account of this signaling system, see van der Vorst 2014).

Metaphor, as we've reiterated throughout this book, plays a vital cultural role in shaping how we ascribe value to and hierarchize the senses. Body-related metaphors and other turns of phrase are important, if not always apparent to us, in how we interpret and understand sense perception. Those involving the ear, hearing and listening abound within the English language and are a crucial aspect of how, linguistically, we orient ourselves to others and make sense of the world around us, and of how we come to construct "ontologies and epistemologies of the acoustic" (Ochoa Gautier 2014: 3). Ears, hearing and listening, for instance, all figure prominently in figures of speech that seek to convey intimacy and a confiding voice ("may I have a word in your ear?"), trueness of character ("her words rang true") or good news ("the words were music to his ears"). They are also used to convey comprehension and understanding ("I hear you"), its opposite ("it's all just noise to me"), as well as a wider sense of socio-political awareness ("the world is listening"). Listening as an acquired skill is also captured in the phrase "to play it by ear," which suggests acting spontaneously and according to the situation, based on an ability to improvise; while a catchy tune is described as an "earworm" (from *Ohrwurm*, the German word for earwig). The figure of the ear also takes on especially complicated metaphorical significance in the works of William Shakespeare (Wang and Tian 2022; Fabiny

2005). More broadly, embodiment-related terminology infuses the language we draw on when referring to mobile devices and their accoutrements, as we see with *handy* (the German word for mobile phones) and, in the case of hearing, *auriculares* (the Spanish word for earphones).

Hearing is generally thought of as a standalone sense, with its own distinct and distinctive sensory capacities. In practice, however, our senses intersect and overlap, working synesthetically in obvious and less obvious ways. For example, hearing, as we explore in greater detail below, can perform a vital preemptive function in orientating vision, such as when seeking the source of a loud bang. Walking and hearing also combine in important ways, ranging from waiting for the sonic notifications at pedestrian lights that indicate it is safe to cross (especially for vision-impaired pedestrians), to the phenomenological encounters prompted by "sound-walking," the practice of deliberative listening "while moving through a place at a walking pace" (McCartney 2014: 212). Neuroscientists have also begun to rethink distinctions between hearing and touch, suggesting that there is significant "crosstalk" between these two "modalities," and that "frequency channels are perceptually linked across audition [hearing] and touch" (Yau et al. 2009: 561).

Listening Phenomenologically

Hearing relates to the physical process of receiving sounds in the ear; listening involves interpreting, understanding and responding to received sounds. When we ponder the question "what is it to listen phenomenologically?"

– which Ihde has done at length in *Listening and Voice: Phenomenologies of Sound* (2007) – it becomes clearer how our sense of hearing both interacts with our sense of sight (in particular), and how sound is differentiated from sight; that is, what is unique about the sense of hearing and the act of listening.

In terms of their interaction, we might, as noted above, think of how sounds often provide anticipatory auditory cues for sight. To cite an example local to us authors, when in the Australian bush, one is quite likely to hear the loud crack of a large eucalyptus branch before one sees it crash to the ground. Or, to repeat Ihde's examples, a bird-watcher in the woods will often first hear a bird and then seek and fix it in the sights of their binoculars, or a person who drops a nail while hanging a picture will know where to search for it due to the sound it makes as it hits and bounces across the floor before they see it.

Just as sight forms part of a larger visual field, sound can similarly be understood, in phenomenological terms, as occurring within a broader auditory field (Ihde 2007). While the two (the visual and the auditory) obviously overlap and interact, the auditory field has several dimensions to it that are quite distinctive. At a basic level, for instance, sound, unlike sight, can penetrate mass: we can hear others speaking through a solid wall but can't see them; and we can tap a cavity wall in order to locate a stud, which we can't see, in order to hammer a nail into it on which to hang a painting. What is more, when it comes to bodily perception of sound, we *hear* with our bodies: we are able to sense reverberations and vibrations in and through our bodies – as much as we do with our ears – such as when we "hear" the arrival of a received text message through the vibration against our body of the phone in our pocket. It is for this reason that Ihde casts ears as the "*focal* organs of hearing," rather than the sole means by which we hear. Indeed, for a hearing-impaired person, "the feeling of the body of sounds" becomes the "focus" (Ihde 2007: 44).

The auditory field is also uniquely pervasive (and invasive) in ways that the visual field is not. This becomes evident when we think about the importance of sound when we are trying to concentrate. If we have a distracting sight before us, we can at least close our eyes; it is often much more difficult to screen out auditory intrusions. Our ears are always "open." For example, as Rowan tries to concentrate on writing this paragraph, he can hear a child screaming in a nearby park, his son playing piano upstairs, another of his children slamming a door elsewhere upstairs, and a Harley Davidson motorbike starting up just as an empty dump truck rumbles past the house. When we begin to take note of our listening, as Rowan just did, we begin to notice that the auditory field is not necessarily continuous in the way the visual field is; it is made up of fluctuating sounds and different, often competing, temporal rhythms (Lefebvre 2004) and disturbances: at certain times, sounds co-exist and overlap; at others, they succeed each other, surging and then falling away. There is an "inconstancy" (Ihde 2007: 57) to the auditory field that we don't generally encounter with the visual field.

Despite this, when we engage with the richness of the auditory realm, sounds, rather than being amorphous or vague, are often isolated and understood as the sounds of *things*. In Melbourne, for instance, an unsighted bird taking flight is readily identified as a crested pigeon due the unique alarm sound it creates with its primary flight feathers (ANU TV 2017) (a sound, incidentally, not unlike the high-pitched squeal of a faulty serpentine belt in a car engine). It is perhaps for this reason that specific sounds not readily identified as known "things" can be either a source of curiosity or a cause for alarm. Thus, despite the distractions of the auditory field, listening routinely involves countless tasks and acts of listening "that call for great accuracy and discrimination": we can clearly hear a buzzing mosquito, often before we see it, despite the insect producing "only one-quadrillionth of a watt of power", while a trained mechanic can isolate a specific mechanical issue or can tune an engine just by listening to the sounds it makes, and a skilled musician can identify a single mistimed note in the midst of a lengthy multi-instrumental performance (Ihde 2007: 7).

What is more, at an experiential level, we can also determine certain "shape-aspects" (Ihde 2007: 62) to things, even when we are unable to see them. For instance, one can distinguish the sound of objects rolling inside a closed box as that of a die or pair of dice, or as that of a hexagonal pencil, based on the sounds they make against the interior surface of the box (Ihde 1982). Here, "the very texture and composition as well as the shape-aspect is presented in the complex richness of the event" (Ihde 2007: 62) of them rolling within an enclosed container.

A key focus of this book is body–technology relations and, in particular, how our "corporeal schema" (Merleau-Ponty 2004 [1962]: 164) is shaped by our habitual engagement with tools and technologies. In Merleau-Ponty's famous example of the sight-impaired man and his walking cane, mentioned elsewhere in this book, he notes how, through habitual use, "the stick is no longer an object perceived by the blind man, but an instrument *with* which he perceives" (2004 [1962]: 176). In Ihde's discussion of this passage, he notes how the sight-impaired man not only feels the sidewalk through his walking stick, he *hears* it, too, insofar as the sidewalk surface takes on an "auditory surface-aspect" (2007: 68). We might discuss the physician's habitual use of the stethoscope in similar terms, or an audiophile's embrace of bone-conduction headphones that touch and convey sound to the inner ear by touching bones behind the ear rather than by enveloping or being inserted into the ear.

In examining these body–technology relations, we seek to explore what is "dilated" (opened up, expanded) and what is "contracted" (reduced, closed down) through our use of mobile phones and other mobile media devices as tools for hearing and being heard. The very transportability of the cell phone and other portable media devices is seen to reconfigure socio-spatial relations in a number of profound ways. The mobile phone is often described as a personal, intimate, private medium (Kuipers and Bell 2018; Hjorth 2009). One explanation for this is that the mobile telephone brings the private and the domestic into public view, without ever being entirely untethered from the domestic (Morley 2003). In this way, the mobile reconfigures the public and the private: formerly private conversations can be had on the move, but these are also opened up to public scrutiny – private conversations can be had in public but are also heard by a listening public. This has necessitated negotiation around the social etiquette associated with the public use of mobile phones (Campbell 2008;

Srivastava 2005; Monk et al. 2004) – something that also had to be established when the fixed-line phone was introduced (Fischer 1992) – and has led to increased awareness of issues relating to listening and privacy management (Worthington et al. 2011). The "always on" (Baron 2010) availability of the mobile phone requires contextual negotiations around where, when and whether or not to take a call. As we shall explore below, these negotiations fostered inventive means of responding to this "crisis of the summons" (Licoppe 2010a), including through the use of once-popular musical ringtones (Gopinath 2013). And, finally, the arrival of personal, music-related, portable media devices, such as the Walkman and (later) mp3 players, prompted strong debate around the extent to which these had deleterious impacts on the social and spatial life of the city.

On the one hand, it has been argued that the Walkman "contracts," in that it "negates" chance urban encounters (Williamson 1986: 210) by erecting an enclosed, auditory wall of sorts between the user and "the geographical space of experience" (Bull 2004b: 112). The Walkman has been described as "the ultimate object for private listening" (Hosokawa 1984: 168): "Walkman users appear to achieve a subjective sense of public invisibility. The users essentially disappear as an interacting subject, withdrawing into their chosen privatized and mobile states" (Bull 2004b: 112). On the other hand, it is suggested that the situation is rather more complicated than this. Connor writes that the Walkman user "is not withdrawn from the scene they're walking through." Rather, they are "often creating a kind of chance collage between the sounds that are filtering through and partly contingent and the organized sound they're hearing" (1999: 308; for a valuable overview of these early debates, see Ferguson 2008: 69–73). Writing on the mp3, Simun develops related arguments, suggesting that "users choose the *degree* of attendance and presence they grant to the places they navigate" and, in some cases, "reconfiguring their environments with music can actually make them *more* receptive to their surroundings" (2009: 931, 932).

In this chapter, we explore body–technology relations and embodied acts of hearing with the aim of providing an account of the "richness of listening" (Ihde 2007: 8) in specific relation to our contemporary bodily engagements with mobile media. In particular, we discuss the relation between embodiment and "telepresence" (presence at a distance) in the context of mobile media interfaces, focusing specifically on the embodiment-telepresence relation in terms of the *aural* effects afforded by mobile phone technologies. Here we suggest that attending to the "soundings" of mobile phone use reveals complexities and modes of contemporary embodiment not adequately captured by the visual primacy, discussed earlier in the book, that is frequently given to the relation between screen-based media and (tele)presence. Moreover, while the telepresent effects of digital communication technologies have been much discussed, they are rarely discussed in relation to the auditory. What is more, telepresence is often used as a generic term that covers a wide range of interfaces and mediated communication, such as the telephone, television, radio and web-based interaction. In contrast, in this chapter, we begin with the premise that each of these media educes a particular telepresent modality and a particular way of being embodied and in-the-world.

Mobile Soundscapes

In *The Tuning of the World* (1977), Schafer examined the cultural specificity of contemporary sound and its effect on our collective behavior; he used the term soundscape "to describe the total experienced acoustic environment. This included all noises, musical, natural and technological" (cited in Bull 2004a: 189). As Bull convincingly argues in *Sound Moves*, portable sound-based technologies such as the Walkman, the iPod and mp3 player, and the cell phone (and since then, smartphone) have contributed (along with the automobile) to the transformation of the urban soundscape by way of an "auditory privatization" of public space (Bull 2007: 32). That is, such devices allow the user to control sound in order "to manage and orchestrate their spaces of habitation" (Bull 2004c: 283). Bull writes:

> Technology has empowered the ears – it has turned the ears from the most democratic of the senses ... to the most exclusive. This empowerment is embodied in earphones, which supplant the uncontrollable and chaotic noise of the street with the chosen sounds of the individual consumer. The price of technologically mediated empowerment is privatization. (Bull 2007: 21)

For example, Rowan manages and orchestrates the auditory space of the train by putting on his noise-cancelling headphones and carefully selecting music to play, thereby creating a "cocooning effect" and setting a certain mood that will mean he can avoid distraction and focus on other things. Only then does he switch sensory modes, moving between audio and visual and textual media – checking emails, scrolling through social media feeds, watching videos, and so on. On other train lines, auditory expectations are addressed (if not fully resolved) through the creation of "quiet carriages," where phone calls, loud music or raucous conversation are discouraged.

Yet, though it can be argued that handheld and attachable technologies such as mobile phones and other portable devices (like mp3 players) have effected a kind of privatization and transformation of the acoustic environment, their sensory and communicative impact diverges considerably. For while the iPod or mp3 player provides a continuous sound-bubble, or "sonorous envelope" (Bull 2004a: 185), that effectively allows the user to deny the contingencies of the outside world (189), the mobile phone is experientially discontinuous, "puncturing" time and space via the sporadic and unpredictable contingency of unexpected calls and text messages or the playing aloud of other forms of audio. For the high school student who plays TikTok videos through their phone out loud in a train carriage, either to themselves or to their peers, these videos might provide entertainment, sonic comfort or, as we explore below, a means of "signing space." For other passengers, the playing aloud of these videos might appear to be done with little apparent understanding of or concern for noise, context, questions of personal taste, or the wishes of other travelers. The sound of these videos is imposed upon the listener and can appear intrusive, lacking choice. In this way, sound and smell are similar in that both are invasive in ways that sight is not. Sound and smell permeate enclosed, public spaces like train carriages and are difficult to "switch off" (with sight we can, at minimum, look away or close our eyes). The mobile music player (such as the once-popular mp3 player or other device where music is played through ear or headphones) is thus discrete and cocooned, whereas the mobile phone user "colonizes" urban space, intermittently carving out a space of communication and telepresent intimacy, temporarily irrupting their immediate soundscape with personal ringtones, bleeps, one-sided conversations and (sometimes) music or other forms of sound.

This colonization often requires a complex negotiation of public and private physical and auditory space. Our pedestrian trajectories can be quite radically revised and re-possibilized by the interruption of a mobile phone call or text message, or by those telepresent on the other end of the phone becoming "virtually" integrated into one's route (through walking Zoom or Teams meetings), or effecting a change to one's route ("can you pick up some milk on your way?").

To illustrate these complex negotiations around the public and private, and physical and auditory space, we draw on three cases from the established mobile phone literature. In the first case, Light provides a number of examples of in-the-moment decisions London commuters have to routinely make when managing phone-related interruptions while on the move. One of Light's participants describes their attempt to manage the arrival of an ill-timed phone call while negotiating their exit from the upper level of a double-decker bus without missing their stop (2009: 201–2). For another, a pleasant walk to the station while listening to music on their mp3 is interrupted by the "obnoxious, jarring" ring of an incoming call (cited in Light 2009: 205); this individual was also concerned to end their call before reaching the train station as they did not feel they would be "able to focus on the call and [other] people moving around ... while ... trying to walk very fast" (cited in Light 2009: 206). In the second case, Licoppe (2010a, 2010b) examines how French mobile phone users respond to the "crisis of the summons" (Licoppe 2010a) – "the pressure of 'accepting a call'" (Ronell 1989: 13) – through the tactical use of musical ringtones. The study reveals two main strategies in the use of the ringtone as a mechanism for managing interruptions. The first involves "weakening the summative force of the ring." This is achieved by replacing the stock call sound with a musical melody as a means of achieving "a short-lived compensatory pleasure" that is traded against taking the call. The second involves "personalizing the sound extract with respect to the callers" so that one or more contacts in a phone user's address book are given specific ringtones. The idea is that these help the user determine whether or not to take a call by providing prior knowledge of the caller's identity, even though this second approach may in fact exacerbate the "crisis of the summons" (Licoppe 2010a: 294). In the third case, Lasen examines the relationship of the public and private, and physical and auditory space, as they are negotiated by mobile phone users in Madrid who like to play music out loud. Lasen writes: "By playing music loudly on their mobiles, mixing the music they love with the sounds of the places they are in, people share and sign their music and their specific way of listening to it" (2018: 107). Their playing of "disruptive ambient music" out loud, Lasen suggests, also becomes for them "a way of signing a space, of signaling a personal territory within the city where they move" (96). The result, however, is a "territorial paradox": it can be comforting to the mobile phone user while in public, and open up opportunities for exchange and interaction, but it also carries the "possibility of upsetting current expectations and formal and informal norms of appropriate behavior in public" (Lasen 2018: 107).

In light of these three cases, we note the distinction Helyer (2007) makes between "intrusive" and "implosive" audio. Intrusive audio describes (for example) the invasive sound of a portable transistor radio (Wall and Webber 2014) or a beat-box (see Schloss and Boyer 2014) – the latter famously used as an auditory focal point and key plot driver in Spike Lee's *Do The Right Thing* (1989). While implosive audio describes the "micro-acoustic-ecologies" of the mp3 player and mobile phone. Yet it is perhaps more accurate to consider the mobile phone as being both intrusive (ringtones, conversations, disruptive ambient music played out loud) and implosive (when using a headset), but also more than this (text messaging while on "silent," and noise-cancelling headphones with "ambient sound" activated, are neither aurally intrusive nor implosive).

Nomadic Telepresence

It is nonetheless the case, as we explore in the later chapter on feet, that both music players and mobile phones are transforming (and continue to transform) our co-proximate and co-present behavior in public places and the way we inhabit and negotiate urban spaces. Indeed, Bull insightfully observes that in mobile privatization there is a desire for proximity, for "mediated presence that shrinks space into something manageable and habitable" (Bull 2004a: 177). The mobile phone in particular offers us the possibility of proximity with familiar others while on the move, and frequently this is via an aural/communicative telepresence (though this is complexified by video-phoning, image/text messaging, and increasingly, the proliferation of web-based made-for-mobile applications). In what follows we consider some of the properties of telepresence, and how the particular conditions of audile telepresence might provide a more complete yet nuanced understanding of this much-used term.

There is no doubt that digital media and communication technologies have irretrievably altered our normalized sense of embodied "location" and "presence." For example, Barthes, as quoted at the outset of this chapter, draws attention to how, in a phenomenological sense, when making a phone call the speaker "collects his whole body in his voice" while the receiver, in the moment of receiving this voice, collects his or her whole self in their ear (Barthes 1991: 252). In twenty-first century "teleculture," it has become received wisdom that it is no longer possible to consider space in terms of the dichotomized categories of here/there, near/far, personal/private, inner/outer or even nowhere/everywhere – dialectics which dominated our understanding at the beginning of the twentieth century. Technological developments ranging from the telephone through to television, cinema, video games and web-based social networking systems have created quasi-spaces where a sense of presence can be felt beyond the location of the physical body. As numerous theorists and philosophers have suggested over the past decade or so, our increasing remote interaction with the world – the possibility of extended intervals of telepresence or telematic perception – indicates a need to rearticulate our collective embodiments, to think through other ontologies, other ways of being-in-the-world and, in a Heideggerian sense, of being-with-equipment (Heidegger 1977). Although telepresence has often been described as a state of quasi-disembodiment – we're often all "hands, ears and eyes" – in a phenomenological sense it is nevertheless still a way of having a body. In general, the body–telemedia relation modifies spatial and sensory perception by changing what is "proximal," or the relation between "here" and "there," into a kind of "distant presence" that becomes part of the "as if" structure of our awareness (*pace* Heidegger). The changed nature of proximal relations becomes apparent, for example, when we observe a person on public transport chuckling to themselves or trying to suppress laughter while watching or listening to some form of humorous content on their smartphones. Ordinarily, if someone nearby is laughing at something, we might either hear what it related to or ask them what they are laughing about. The "mobile privatization" and "distant presence" associated with public use of mobile phones creates changed proximal relations such that we are unlikely to ask such a question.

As noted above, the phenomenological body literally apprehends and appropriates technologies-in-use into an adaptive corporeal schema, and we could argue that it is precisely our capacity for ontic dispersion beyond the physical limits of the body that leaves us open to the embodied or perceptual distraction of telepresencing media interfaces. The term *distraction* – originating from *distrahere*, meaning to pull in different directions – aptly describes how our attention becomes divided when we speak on the phone (Arnold 2003). It suggests that the locus of our perception is distributed between the "here" and "there," such that we can *know* different times and spaces simultaneously, an effect which shifts the boundaries of what "immediacy" is, and how it is defined and experienced. Indeed, if the condition of embodiment is *defined* in terms of the immediacy and locatedness of sensory experience, then such tele-media mutate "any simple unity of body and sensation towards its capacity to gather up several locations at once" (Waldby 1998). Thus we experience at the very least a tripled or para-ontology, residing *in* our tangible locales (lounges, bedrooms, kitchens, chairs, cars, cafés, public transport), *in* the space of the other, and somewhere *in-between* (the non-space where, in Margaret Morse's (1998) terms, people talking on the telephone "meet").

The shift from mobile telecommunications to mobile media (Goggin and Hjorth 2009), enabled by developments in mobile standards (3G to 4G, and now 5G), has also contributed to increased mobile content consumption and has facilitated a heightened sense of being immersed in something, of being transported somewhere – a sense of "absent bodies" (Leder 1990) unaware of, or bracketing out, their surroundings. For Rowan, this is observed in the habits of fellow daily train commuters. For example, a tall familiar stranger enters the same carriage from the same stop each day. Due to his height, and the narrow gap between seats, he sits at right angles to the seat with his legs projecting into the aisle. With his elbow propped on his knee, he holds his phone horizontally between thumb and forefinger. From the moment he is seated and for the remainder of the forty-odd-minute train journey, he becomes fully immersed in the streamed video content on his phone, and oblivious to those moving about and standing around him.

Despite the close relation between the telephone and the experience of telepresence (Morse (1998) suggested over two decades ago that talking on the telephone described the archetypal moment of telepresence), the latter has more commonly been described by way of metaphors of vision and seeing. In an analysis of the telepresent landscape, for example, Campanella suggests that "[t]elepresence is reciprocal, involving both the observer and the observed. In other words, the observer is telepresent in the remote environment, and the observed environment is telepresent in the physical space in which the observer is viewing the scene" (2000: 27). Yet telepresence can be used to explain the adaptive and reciprocal interaction with any media where we experience visual, aural and/or multi-sensory presence at a place where we are not physically located, when we *extend* our senses *into* a remote location, and simultaneously seize the remote location into our own immediacy. Nevertheless, this very general understanding of telepresence needs further distinction – televisual telepresence is not perceptually or spatially the same as audile telepresence, and neither is the immersive telepresence of video games analogous to communicative telepresence.

As Connor observes, the distinctive feature of auditory experience is its capacity to "disintegrate and reconfigure space" (2004b: 58). In their study of listening, corporeality and presence in real and artificial environments, Turner et al. (2007) also comment on the unique qualities of auditory space. Unlike pictorial or televisual space, they argue, auditory space is not "boxed-in" or framed as a window-on-the-world;

instead, it is fluid and dynamic, and filtered rather than contained or "stopped" by material obstacles such as walls and corners. Thus it is not enclosed or "held" in place, but creates "its own dimensions moment by moment" (Carpenter, cited in Turner et al. 2007). Recent advancements in sound technology, led by Apple and others, have contributed to the development of "spatial audio," audio that is "engineered to have a 3D quality to it" (Lacoma and Cohen 2022), or what has been described as "AR for your ears" (Goode 2021). Phenomenologically, this can take some aural adjustment. For example, when we are listening to music through headphones, our ability to detect subtle distinctions in sound direction can create momentary disconcertion or auditory confusion as to the source of a sound when there are complicated or dramatic sonic shifts from one headphone cup to the other.

It is this elasticity and pervasiveness ascribable to sound that in turn educed a particular way of experiencing the audile telepresence and sense perception of landline telephony. As it came into common usage, the early landline telephone, Connor argues, allowed our interior body – the inner ear – to be pervaded "*almost without mediation*" by the "vocal body of the other" (Connor 2004b: 56, our emphasis). Thus with telephonic telepresence we were afforded the impression of being corporeally tethered to another's body in real time, and thus paradoxically co-present at-a-distance. Connor writes:

> The telephone uses the principle of electromagnetic induction to translate sound vibrations into fluctuations of electrical charge, which are then translated back into movement at the other end. It is the capacity of electrical

> impulses to be transmitted at long distances without significant degradation by and into noise that accounts for the illusion of bodily presence, the sense that the voice that arrived at the other end of the line had not been transported so much as stretched out. (Connor 2004a: 159)

As Hammer (2007) comments, early in the development of landline telephony it was thought that the device could act as an instrument of medical diagnosis, somewhat like a stethoscope stretched along a very long wire.

Yet although the fixed landline telephone was one of our earliest experiences of what Helyer calls "schizophonic" audio – allowing us, as it were, to hear non-present voices – it was still the case that each correspondent was fixed in a particular geo-spatial location, so "the telephonic act became a sonic bridge between familiar sites" (Helyer 2007). With the mobile phone, because the location of the caller is often unknown, shifting or unpredictable, communication becomes "de-territorialized." That is, the particular consequence of the mobile phone upon aural telepresence was an unfixing from place, effecting the emergence of a kind of *nomadic telepresence* (Helyer 2007) that is only partially overcome by asking that most common of questions in mobile communication – "Where are you?" (for discussion of this deceptively simple question, see Wilken and Goggin 2012: 15–17).

Paradoxically, because of this unhinging of one's geo-spatial location, the itinerant attributes of mobile phone conversation often result in a sense of discomfiture or ill-fit between the conversation-context and the very public spaces (buses, trains, sidewalks, etc.) wherein we find ourselves "on the mobile":

> Certain conversations can induce emotional and bodily responses which may be quite incompatible with [mobile users'] perceptions of their physical location. Their participants often look as though they don't quite know what to do with themselves, how to reconfigure the tones of voice and postures which would normally accompany such conversations. The mobile requires its users to manage the intersection of the real present and the conversational present in a manner that is mindful of both. (Plant 2003: 50)

This awkwardness is perhaps a result of engaging in telepresent communication in public places, of having to negotiate between "presents" (telepresence and co-presence), because mobile phone communication when one is on the move and on the street is rarely immersive and private, but demands an awareness of both others and one's immediate surroundings. In such contexts, use of mobile phones in public can afford a mode of "listening in" that works to blur the boundaries between the public and the private. At times, this can seem voyeuristic, almost transgressive. As listeners, we are privy to conversations, scenarios and personal details that we might otherwise not gain access to. When these private phone conversations are conducted in enclosed public (or public-private) spaces, such as a café or train carriage, it can be difficult to block them out or (to mix modes) to "look away"; they are just there. When this occurs it can feel like we are almost forced to listen to these conversations, like in the rather uncomfortable recent experience one of us had in a café when another patron began describing a medical procedure in some detail to the person they were speaking to on the phone. Ear or headphones afford a means of blocking out such conversations, but we don't always have that luxury.

In a relational sense, mobile technologies are being used not just to "compensate for the absence of our close ones"; rather, they "are exploited to provide a pattern of mediated interactions that combine into 'connected relationships,' in which the boundaries between absence and presence eventually get blurred" (Licoppe 2004: 136; see also, Farman 2012). In an embodiment sense, mobile technologies require us to adeptly shift between actual and telepresent space and to "behave" in ways that accord with (or deviate from) conventional modes of being-on-the-phone. Plant (2003) developed a typology of mobile phone behaviors, including the sometimes tricky footwork involved in stopping or diverting one's pedestrian trajectory, and the "space-making" practices of bowing the head, or shielding one's mouth or face with the hand to mark out a temporarily private aural space (see the chapter on feet for further, detailed discussion). The body thus undergoes the aural and spatial disciplining required by the device, the collective mobile-user habits of a culture, and the demands of telepresent communication in public places. What has emerged over the past few decades, then, is a dynamic corporeal schematics of mobile phone use, and an ongoing and adaptive management of audile telepresence in the urban soundscape.

These negotiations, it needs to be acknowledged, are made significantly more complicated for the Deaf and Hard of Hearing. For the Deaf and Hard of Hearing, overlaying the situational complexities attending the demands of telepresent communication in public places are structural and other challenges relating to "audism" – an underlying "audiocentric privilege" that discriminates against the Deaf and Hard of Hearing (Eckert and Rowley 2013). While sensors embedded in smartphones hold promise as emergent assistive technologies, and "rejoin and remediate a long history of devices aimed to support" those with hearing and other disabilities (Goggin 2016: 537; see also Maiorana-Basas and Pagliaro 2014), many of the complications for the Deaf and Hard and Hearing are played out at the level of "everyday data cultures" (Burgess et al. 2022) and daily communicative practice. On the one hand, this may involve negotiating questions of "passing," which is described as "the way people conceal social markers of impairment to avoid the stigma of disability" (Brune and Wilson 2013: 1): "Can I pass as hearing? Shall I even try?" (Bitman and John 2019: 64; see also, Harmon 2013). It has been noted that "accessible features of smartphones, and their pairing with assistive technologies (e.g., cochlear implants, Telecoil position on hearing aids) are crucial for passing and the avoidance of stigma" (Bitman and John 2019: 66). On the other hand, it may also involve negotiations around desired or preferred modes of mobile communication. In their study of Israeli Deaf and Hard of Hearing smartphones users, Bitman and John report that, "despite perceptions of video calls as culturally preferred by Deaf users," their interviewees "reported feeling uncomfortable with the medium," often preferring to use WhatsApp for text-based communication (2019: 63). Even so, the felt pressure to respond to calls – Licoppe's (2010a) "crisis of the summons" – is particularly acute for the Deaf and Hard of Hearing. Bitman and John record that "even smartphone users who are not auditorily able to perform voice calls still experienced strong pressure to comply with its norms of communication." In this way, they observe, "the mobile voice call becomes a mechanism of oppression" (2019: 67).

Mobile Listening: The Smartphone as "Ear"

We argue throughout this book that our body–technology relations and our "corporeal schema" (Merleau-Ponty 2004 [1962]: 164) are shaped by our habitual engagement with tools and technologies. In an auditory context, for instance, it has become commonplace to regard the hearing aid as an auditory extension of our body. Once incorporated within our corporeal schema, the hearing aid "withdraws" and is thereafter "barely noticed, if at all" (Ihde 2010: 136). (The cochlear implant, however, is understood to present a more complicated case, as "first-hand accounts from cochlear implant users overwhelmingly emphasize constrained psychotechnical experience and ongoing negotiations with body–technology compatibility" (Mills 2014: 263).)

Curiously, it is comparatively less commonplace for the smartphone to be regarded in similar terms as an auditory extension of our body, despite evidence to the contrary. To begin with, Crawford notes that the contemporary smartphone "contains a conglomeration of sound technologies," including inbuilt microphone, speaker, headphones or earbuds; it also serves as a platform for various forms of sound-related software (2012: 14). She goes on to detail four ways of listening with a smartphone. The first involves listening *in place* (through the use of the phone as a music-playing device) and listening *to place* (through apps like Shazam that utilize the phone's inbuilt microphone to act as an external "ear" that listens for and identifies music being played in that space) (2012: 214–18). The second is what Crawford terms "network listening" (218); that is, using the smartphone as a platform for social media apps, with the phone serving as "a significant agent in the ability to engage in background and reciprocal listening throughout the day, untethering the modes of social media listening from desktop environments and allowing for potentially ongoing attentiveness, regardless of location or context" (219). The third, "biometric" listening, enrolls smartphones (and other portable devices like Fitbits and smart watches) in "ever more intricate forms of self-listening and self-management" in support of "an increasing reflexivity in the relationship between users and their mobile phones" (220). This can take a number of forms and can serve a variety of purposes. There are, for instance, self-management "white noise" apps, which come with various color settings (from pink to brown to blue, with each color referencing a different frequency), that, when listened to, create an auditory "mask," making sleep or concentration easier. There are also meditation-related apps, such as Calm Sleep and Meditation, that provide material to listen to while also "listening" to the user's biometric signals. The success (or otherwise) of such apps can vary between individual users. In addition, there are also tactical uses of biometric listening services. For instance, one study examines how Turkish housewives have made use of smart watch data, such as the number of steps taken each day, to emphasize to family and friends just how demanding unpaid domestic labor can be (Özkan, forthcoming). The fourth form of mobile listening Crawford terms "eavesdropping" (223), where the smartphone serves as a "listening" (tracking) device, gathering latitude and longitude and vast amounts of other traces of our digital passage through time and space. For Andrejevic, the last of these forms of listening is why the mobile phone plays a central role in embedding an automated "drone logic" into everyday life. The idea here is that the smartphone, as a drone-like probe, works to reinforce an "emerging logic of portable, always-on, distributed, ubiquitous, and automated information capture" (Andrejevic 2015: 195). In this way, the mobile phone "extend[s] the reach of the senses and add[s] new capacities" (McCosker and Wilken 2020: 63), especially auditory capacities.

Conclusion

In *Understanding Media*, McLuhan famously suggested that we should consider the telephone as an extension of the ear (McLuhan 2003 [1964]: 289). Taking up this suggestion, in this chapter we have explored the ear–mobile device coupling and the particular sensory modality of hearing. We first traced the biology of hearing, and the multistabilities and phenomenology of hearing and listening. We explored histories of mobile device use and practices of listening and aural/oral connection. We then turned to consider typical scenarios of use, such as the "immediacy" of interpersonal conversations, the "out-of-time" privatization of listening and the "cocooning" of the self via music and podcasts, how mobiles have transformed (and continue to transform) our co-proximate and co-present behavior in public places, and the way we inhabit and negotiate urban spaces, before touching on how the mobile medium itself operates as a "listening" device that captures individual micro-data. In this chapter, we have suggested that the "sensing" of mobile phone communication elicits an intimately audio as well as visual, haptic, "handy" and visceral awareness, as a mode of embodiment that demands the ontological coincidence of distance and closeness, and presence and telepresence. In particular, we have sought to re-think the phenomenology of the mobile by giving emphasis to its auditory capacities. We have argued that by considering the intersensorial condition of our being-in-the-world, discerning the sensory "complexion" of mobile phone use, and attending specifically to the corporealization of sound and aural perception in the context of mobile media, we are equipped with a more "full-bodied" and extensive interpretation of the embodiment–telepresence relation in contemporary urban spaces.

In the next chapter, we consider hands and tactility and the central role they play in our contemporary engagements with mobile devices. The chapter is thus framed around an examination of haptics and of the hand as a world-shaping tool.

4
Hands

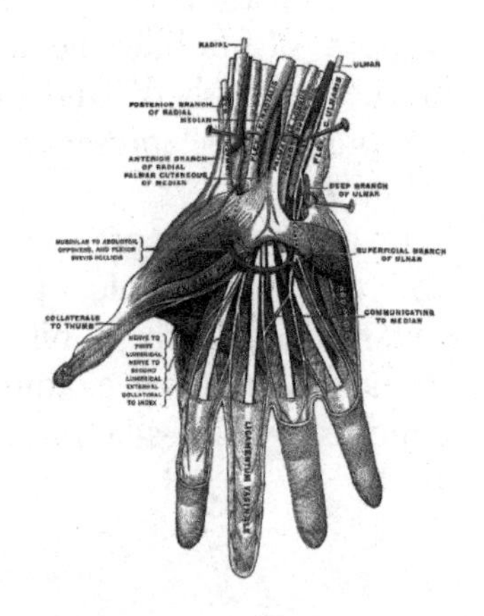

The hand which scoops up water is the first vessel. The fingers of both hands intertwined are the first basket ... It is not enough that this or that shape should exist in the surrounding world. Before we could create it ourselves, our hands and fingers had to enact it ... Words and objects are accordingly emanations and products of a single unified experience: representation by means of the hands. (Canetti 1962: 217)

The hand is a world-shaping tool and our primary instrument of touch, arguably the most intimate and active of the senses. The significance of the hand can be traced along numerous trajectories. In anthropological and sociocultural terms, it has been crucial to our material survival through the collaborative shaping of equipment and environment, while at the same time vital to personal, intercultural and cross-cultural communication; consider, for example, the non-verbal universality of many gestures, the "handiwork" required for artistic and creative expression across cultures, and the ways we enact care and love for humans, animals and objects with the caress of our hands. In biomechanical and anatomical terms, the hand is a dexterous and agile manipulator, and one of the most elaborate and intricate features of the human sensorimotor system. With twenty-seven individual bones – eight carpal bones, five metacarpal bones and fourteen finger bones (also called phalanges) – connected by many joints, muscles and ligaments, including nineteen links that enable twenty-four degrees of freedom, hands are capable of learning and performing a complex range of grasping, pinching and load-bearing functions (Jaworski and Karpiński 2017). In neurophysiological terms, hand movements and sensations occupy large areas of the cerebral cortex, including receptors for vibration, texture, temperature, slippage and stretching. The hands are "innervated" by at least fourteen different types of nerve fiber, all encoding different stimuli and capable of detecting gaps and elevations as small as one tenth of a millimeter (Jones 2006).

The opening quotation from Canetti aptly captures the sweeping impact of the hand on human life at both material and communicative levels. In his detailed study of its neurological, linguistic and cultural significance, Wilson goes even further:

> The hand is so widely represented in the brain, the hand's neurologic and biomechanical elements are so prone to spontaneous interaction and reorganization, and the motivations and efforts which give rise to individual use of the hand are so deeply and widely rooted, that we must admit we are trying to explain a basic imperative of human life. (Wilson 1998: 9)

We also distinguish between "discriminative touch" – that which gives us facts about location, movement and strength – and "emotional touch" transmitted by C tactile fibers that provide information to the brain crucial for social bonding (Stromberg 2015). Each of these trajectories and capacities – anthropological, biomechanical and neurological – are of course not discrete but tangled in the long and deep histories of hands and their role in the coevolution of society and sensoria. Of more relevance to our focus in this chapter, and the book more broadly, there is no doubt that, phenomenologically, hand–tool relations are among the most noteworthy and formative for our individual and collective body schemas and habits.

In this chapter we explore how the hands and touch – the tactile elements of sense – are central to our contemporary engagements with mobile haptic devices. In becoming an incorporative aspect of the hand, mobile touchscreens – as we navigate them with our fingers and learn new gestural literacies (swipes, flicks and pinches) – have entered into an intimate and habitual relationship with a body part that is in itself of some consequence as a tool and conduit of communication and worlding. As Lasen (2004) insightfully commented at the beginning of the mobile phone era, our relationship with handheld devices is unique because they are continually touched and held close to the body, such that we become intimately attached to them both instrumentally and emotionally.

Informed by haptic media studies, the corporeal or sensory turn in contemporary research, and phenomenology, this chapter first explores the emergence of haptic and touchscreen technology. We then turn to a discussion on the significance of touch, a feature of human intimacy that has been particularly highlighted in recent times, and consider the tension between touch as a vital sensory modality of human experience and mobile media practices during the coronavirus pandemic, when proximity and (tactile) intimacy with other bodies in urban and domestic spaces became fraught with the risk of viral contagion. In many ways, the pandemic has required us to reconsider how touch relations are conceptualized, enacted and embodied. In this context, we examine the role of haptics and the touchscreen and the possibility they afford us – through the hands and via embodied and material metaphor – to extend our corporeal reach through the mobile interface. While this modality of distant touch might be incomplete or imperfect, we ask if mobile media might play a role in narrowing the sensorial gap by offering an "as-if" mode of experience that is becoming part of our collective habits and repertoires of the hand, supporting the claim that our everyday realities are formed at the nexus of embodied perception and media apparatuses.

The Haptic Interface

Haptic media engage the many facets of touch, a sense ensemble that incorporates cutaneous, kinesthetic, proprioceptive, somatic, mimetic, metaphoric and affective modes of perception (Paterson et al. 2012). In his text *The Senses of Touch: Haptics, Affects and Technologies*, Paterson explores both the historical and contemporary dimensions of touch. He argues that touch cannot simply be defined in physiological terms; it is also always "a sense of communication" (2007: 1), and, more significantly, it is *manifold*, a term that reflects the incredible complexity of tactile sense perception. The turn to "touch theory" is interwoven with the emergence and popularization of haptic technologies and, in media studies, is often applied to the analysis of computer and touchscreen interfaces. As Goggin points out, our current "preoccupation with touch, contact, hand, thumb and finger typing, and device manipulation, touchscreens and so on in the development of mobile phone culture point to a more haptic orientation in the wider cultural experience around the materiality of these devices" (Goggin 2017: 1568).

In a detailed media archaeology that traces the historical evolution of the relationship between touching and media interfaces, Parisi (2009) suggests that advertisements for the Nintendo DS had to teach people that "touching is good" after decades of screen etiquette had impelled us not to handle screen surfaces, a narrative that can be traced through to Apple's customizable Taptic Engine and the Apple Watch. It is important to remember this history when considering the now normalized practice of screen touching. Media practices, and the appropriate bodily repertoires they involve, are generational and cultural; they are taught, untaught, retrained.

It is also worth noting, as Cooley suggests, that when designers and developers boast about the supposedly intuitive "feel" of mobile interfaces, they engage in a rhetorical practice that "collapses a plurality of human hands into the abstract ideal of a universal 'human hand,'" which overrides the myriad other factors of skill, (dis)ability, literacy, age and gender with a biomechanical, ergonomic abstraction (Cooley 2014: 28, 32). Similarly, Parisi suggests that beyond the "mundane" haptics that we experience today in mobile phones, wearables and games, future interfaces will at best be built on normative models that posit an impossibly universal body, and that "[g]iven touch's impossible complexity, any attempt to digitally remake it will be necessarily incomplete and fragmentary" (Parisi 2022). As Wilson argues:

> The problem of understanding what the hand *is* becomes infinitely more complicated, and the inquiry far more difficult to contain, if we try to account for the differences in the way people use their hands, or if we try to understand how individuals acquire skill in the use of their hands. When we connect the hands to real life, in other words, we confront the open-ended and overlapping worlds of sensorimotor and cognitive function and the endless combination of speed, strength, and dexterity seen in individual human skill and performance. We also confront the vagaries of human learning. (Wilson 1998: 9)

The multistabilities of touch and haptic screen practices come sharply to light in the context of disability and the "ableist assumptions" (Walters 2014) underlying normative human–technology relations, emphasizing not only that all media are *material* interfaces that "incorporate the bodies of users" (Mills, cited in Goggin 2017: 1564), but also that bodies are variable in their perceptual and sensory capacities. Goggin (2017) notes that the vibration function – now a mundane aspect of mobile media alerts and notifications – is of particular interest in terms of the way it has evolved as an important "communicative element" that is both heard and felt "in" the body by people with disabilities. The initial release of the iPhone in 2007, with its emphasis on touch capability (that was nevertheless solely dependent on visual perception), disregarded the fact that for visually impaired users older mobile phones with physical keypads were always a form of touch media, and outraged the blind community for the patent "lack of respect for the evolving tactile uses of mobile media" (Goggin 2017: 1571). For such users, the emergence of a flat and smooth screen that incorporated the keypad rendered the device less, not more, sensorially available to touch. Subsequent developments in accessibility and assistive features – such as VoiceOver for iOS5 and TalkBack for Android OS – reveal the many differences in bodily experience and the ways body metaphors vary across that experience.

For example, the Explore by Touch feature for Android allows blind users to touch the screen and "hear what's under your finger via spoken feedback" (Google, cited in Goggin 2017: 1574), shifting the primacy of the metonymic hand–eye connection to one that is exclusively hand–ear, such that app activation and touch-based interaction is translated into words rather than triggering animated images and generic sound effects. A user with motor impairment might rely on their mobile device to stay connected, acutely attuned to the vibrations that provide sensory feedback not afforded by traditional computer or mobile keyboards, or access voice recognition to record messages and search the web. The Apple Watch and Taptic Engine, reviewed favorably in press coverage as heralding "transformational change" in accessibility (Koziol 2015), allows hearing and visually impaired people to navigate urban space through the integration of haptics into the device's map function. The "taptic" function (combining "tap" and "haptic") allows wearers to customize tactile sensations in the form of vibrations at adjustable intensities of strength and duration. One self-described deafblind user relates how the Apple Watch "transformed her ability to navigate around town [...] 12 taps means turn right at the junction or three pairs of two taps means turn left" (Koziol 2015). Instead of relying solely on visual and auditory cues, the interface offers a new dimension of haptic mobile media experience that, for some, is more nuanced and personalized. In terms of widescale use, beyond its functional utility, the Watch also provides a sense of intuitive connectedness that one might argue "completes" the affective affordance of mobile media as an extension of the body. As Goggin insightfully comments, tracing the evolution of mobile media via disability discourse and practices acknowledges the "rich, complexly embodied, sensory histories of media users" (2017: 1575); moreover, it highlights how the design and narratives around mobile haptic technologies are often developed according to dominant "social imaginaries," and that innovation can either "contract" or "dilate" the corporeal schemas of all users, with particular consequences for those bodies that do not conform to the imagined norm.

These significant innovations aside, touchscreens are typically designed with what is called a post-WIMP interface (Jacob et al. 2008). WIMP stands for Window,

Icon, Menu, Pointing device, the standard tools for navigation and control of a desktop computer interface. Mobile touchscreens are specifically designed for use with the finger or fingers for multi-touch sensing, and because the screen is a capacitive touchscreen, it depends on electrical conductivity that can only be provided by bare skin and the moisture of the fingertips (even knuckles don't work quite as well). Post-WIMP interfaces require an embodied understanding of – a kinesthetic familiarity with – naïve physics; for example, primary bodily sensations such as inertia and springiness can be found in many mobile applications and games and provide the synesthetic illusion that windows, objects and icons on the device have mass (Jacob et al. 2008). Naïve physics can also include our body-memory of hardware such as the buttons and keyboards that are simulated in the mobile touchscreen user interface. As Pallasmaa (2009: 118) points out in his book *The Thinking Hand*, hand skills are "primarily a matter of embodied muscular mimesis acquired through practice rather than conceptual or verbalized instruction," a fact that is exemplified in many of our media practices. Gamers, for example, typically accumulate embodied memories of game controllers, which become imprinted on their sensoria through the repetition of particular hand actions (and effects), such that a "haptic bond" is established with the interface (Parisi 2015: 17–18). This accretion of sensory memory in the body applies equally to the mobile touchscreen.

Thus, there is a certain haptic intimacy that renders the screen an object of tactile and kinesthetic familiarity, a sensory knowledge that accrues in the fingers to correlate with what appears on the small screen. According to Eikenes and Morrison (2010), we have developed a new kind of "motion literacy" specific to devices that combine touchscreen with accelerometer or position-recognition functionality. A type of kinetic and motile learning is required – or, in phenomenological terms, the absorption of a "fresh instrument" into our body schema – that works to overcome or adapt to the imprecise control we have over objects and actions on the screen. The hand becomes quite literally a physical metonym – a material metaphor – for the body, allowing us to conjoin "two different orders of reality, real physical gesture and its on-screen representation," such that we experience a "kinetic materiality" that corresponds to the action and movement that takes place on the screen, creating moments of tangibility and concreteness (Rush 2011). In part, this is achieved by kinesic natural mapping (Skalski et al. 2011), an effect enabled by the way touchscreens deploy physical analogies. Natural mapping works to "complete" being in a mediated space, facilitating an immersive experience. This is possible because of our ability to take on an "as if" structure of embodiment, a concept that we expand upon below.

These new skills, accommodating naïve physics, and kinetic and motion literacies, are nowhere better exemplified than in mobile phone games that depend on reality-based interaction and enfold the player into a temporary and incomplete simulation of real-world physics, like turning a page in the origami-like puzzle game *Tengami*, or "finger-bombing" in *Angry Birds* (Parisi 2008). Here the kinetic experiences of tangibility, concreteness and elasticity become "condensed in the hand" (Kirkpatrick 2009: 134). An overwhelming number of games and apps available on mobile phones and tablets are mimetic to some degree; that is, they simulate embodied experience (such as moving a block or sling-shotting a bird) and, ideally, are designed so that they can be used immediately and intuitively rather than requiring instruction. As Keogh notes, mimetic interfaces call upon bodily habits and memories that "more closely align to those already learned through the player's everyday existence" (2018: 64). The affordances of the haptic mobile interface – intimate, tappable, pinchable and intuitive – combined with its portability, have had a significant impact on game design and development, and rendered games mimetic and thus accessible to everyone.

Throughout research conducted as part of a three-year project investigating mobile media in the home (Hjorth and Richardson 2020), participants have frequently described the sensory pleasure engendered by the experience of playing mobile mimetic games. One participant described her obsession with *Godus* (a world-builder that enables detailed sculpting of multiple layers of land, with each configuration being unique to the player), explaining that she played the game intensively for over a year: "there was something about the way you could sculpt the land directly with your fingers, and set your little workers to build or mine, that was really satisfying." Such new sensory literacies and pleasures are not just important at the level of embodied attachment to one's device, for the haptic and handy intimacy of mobile games also becomes material to our social and personal intimacies. Another participant recalled and then performed how she would share *Angry Birds* gameplay with her partner in the evenings. This involved them lying together on the couch or bed and passing her iPhone between them (the only rule being that the device had to be relinquished to the other if a "life" was lost). They thus enacted a ritual of closeness and being-together that is bound up both in the simplicity and "swapability" of the game and in the very materiality and co-touchability of the interface. For others, the closeness of shared play was also experienced through a sense of networked co-presence, such as the sense of connection that is realized in playing online games with physically distant friends and family, as if "touching" the same game. In this way, touchscreen gameplay expresses not only a way of being in the world, but also a way of being together through "mediated social touch" (Paterson 2007: 131). In the sections below we further explore this notion of mediated touch, and its capacity to "stand in" for physical touch, particularly in the context of pandemic-inflected mobile media use. But first, we briefly explore the broader significance of touch.

The Significance of Touch

In literal or physiological terms, touch is fundamental to human health. Not only is it seen as especially vital for attachment in newborns and in infant development (Ardiel and Rankin 2010), the importance of touch extends beyond infancy and is necessary for well-being and communication across the life course (Nist et al. 2020). Touching "stimulates pressure sensors under the skin that send messages to the vagus" (Kale 2020), a large nerve that conducts signals between organs and the brain. "As vagal activity increases" as a result of touch, "the nervous system slows down, heart rate and blood pressure decrease, and … brain waves show relaxation." With this, cortisol is decreased, and oxytocin and serotonin are released. In short, "human touch is biologically good for you. Being touched makes humans feel calmer, happier, and more sane" (Kale 2020). Indeed, a loss of touch is detrimental to us as humans, leading to physical and emotional deterioration and developmental delay in children (Ardiel and Rankin 2010). Such studies that seek to document the detrimental effects of touch deprivation are of course premised on the notion of touch as a material, tactile, skin-to-skin experience, a presumption that we challenge in the latter part of this chapter in relation to the cultural and embodied significance of mediated and non-material (or "as-if") forms of touching through the mobile screen.

During the Covid-19 pandemic, risk of contagion from the coronavirus contributed to a widespread loss of physical touch, both in clinical contexts, where it is said to have significant therapeutic benefits (Nist et al. 2020), as well as in broader social and communicative contexts. "From dating to banking, education to retail," Powers and Parisi (2020) noted, "the virus has pushed everyone to rethink how touch and proximity factor into daily interactions." While this loss of touch was most evident and most keenly felt because of the need for physical distancing, it has been suggested that we are in fact experiencing a longer-term "crisis of touch"; neuroscientist Francis McGlone claims that we have demonized touch, and especially social touch, in part due to the potential for litigation and hypervigilance over personal boundaries, and that this has been exacerbated by ubiquitous social and mobile media use, which now mediates interactions that once required bodily proximity or skin-to-skin touch (McGlone, cited in Cocozza 2018). Nevertheless, the pandemic undoubtedly accelerated this culture of no touching. We were separated by Perspex guards and faceshields, disciplined through signs and floor markings to maintain social distancing, and discouraged from touching surfaces (including one's own face) or other people.

Hands in particular were considered pernicious vectors of contagion in need of frequent sterilization, and the mobile phone an unsanitary reservoir of germs transferred from porous hands and fingers. Hands became disciplined: workplaces (including our own) implemented strict regimes around haptic sanitation, with hand wash dispensers placed at the entrances to buildings and new educational signage promoting "correct" ways of washing one's hands added to bathrooms. People were dissuaded from shaking hands and, instead, encouraged to fist bump or touch elbows as gestural modes of greeting – despite the impoverished tactile benefits of the latter. And, in many cafés, restaurants and shops, purchasing with cash was discouraged in favor of contactless payments, which has seen a rise in popularity of mobile payment and digital wallet services. As observed by Powers and Parisi, in a study conducted by the hand-tracking company Ultraleap – a tech development enterprise that seeks to create "the sensation of touch in mid-air" – respondents' concerns about the hygiene of touchscreens were significant enough to prompt "the company to speculate that we are reaching 'the end of the [public] touchscreen era' … Touchscreens are no longer sites of naturalistic, creative interaction, but are now spaces of contagion to be avoided" (Powers and Parisi 2020).

Others have taken a more measured response to the crisis of touch in terms of rethinking the place of mediated touch in sociality. For instance, Jewitt et al. (2021) – a group of computer scientists, haptic designers and developers, engineers and social scientists – developed a "Manifesto for Digital Touch in Crisis," to consider the challenges we currently face and explore how the "complex space of social touch" can be mediated by technologies. As they note, social touch is crucial to our "communicative repertoire" and integral to overall well-being. While social touch (such as hugging or hand-holding) is a multistable phenomenon, differing across cultural habitudes and contexts, it is intrinsic to human corporeality as a way to communicate caring and intimacy, and something that we also share with many mammals such as our companion animals. Jewitt et al. (2021: 1) argue that social distancing practices have disrupted not just the frequency of touch but also *how* we touch, underscoring the need for critical studies into the potential of digital social touch and the rich cultural and perceptual variability of our touch relations, and how this becomes complexly translated into mobile, haptic and touchscreen media practices.

Mediated Touch and the "as-if" Structure of Perception

In phenomenological terms, mobile interfaces are not merely attachments that temporarily enable perceptual reach, they are part of our embodied routines of dwelling and movement. Increasingly, our interpersonal interactions are entangled with mobile media as they become central to everyday life practices and modify our embodied ways of knowing and being with others. This integration occurs by way of deeply familiar material and embodied metaphors that allow us to perceive what Farman (2012) calls *social proprioception.* The term proprioception describes the sense we have of our body's position in space, and of the boundaries of our body parts as we move through space (so we don't bump into things). Farman translates this sense of physical presence into social presence, to refer to our ambient awareness of others perpetually online (primarily through social media), which involves a sensory and spatial "dilation" that is now a habitual and a mundane aspect of our daily lives. That is, our experience of co-presence now routinely includes forms of ambient proprioception or mediated social touch (Paterson 2007), accessed through the handy navigation of mobile screens.

The closeness of telepresence, of shared mobile gameplay and networked co-presence – how we keep "in touch" with friends and family – is now collectively felt; it is a way of being together via mutual spatial and corporeal adjustment. During the pandemic, with closed borders and rolling lockdowns, mobiles came to play an even more vital role in facilitating ambient proprioception, enabling intimate forms of connection at a distance. When used as *telepresent* devices, can personal mobile media effectively "stand in" for or even replace touch? We consider two responses to this question. The first, espoused by both neuroscientists and haptic media scholars, suggests that touchscreen interactions are incapable of replacing "real" social touch or effectively alleviating skin hunger. The second, building on the phenomenological notions of mediated touch and social or ambient proprioception, suggests that mobile touchscreens afford a sense of touch *in potentia*, and that this possibility space manifests because of our perceptual capacity to engage in an "as-if" structure of perception and experience the hand as a metonym or material metaphor of the body. This experience of potential or "as-if" touch, we argue, is importantly *non-trivial*, in that it describes a palpable, felt sense of interpersonal embodied connection during times of physical isolation or distance.

Neuroscientists, computer scientists and haptic media developers and scholars alike have argued that touchscreens – and even more advanced virtual reality technologies – are far from being capable of simulating actual touch and its many benefits. Neuroscientist Alberto Gallace, for example, is highly skeptical of the much-hyped capacity of web-capable touchscreen interfaces to connect us during the pandemic, because they "cannot substitute for skin-on-skin contact" which is so crucial to intimacy and well-being (Gallace, cited in Kale 2020). Others suggest that touch is too complex and multi-modal to be adequately replicated; that is, touch is not a single sense, as different nerve endings "recognize" different sensations, both inside and outside the body (such as itch, vibration, pain, pressure, texture, and even gentle stroking) (McGlone, cited in Cocozza 2018). Moreover, the lived experience of touch is not uniform but nuanced, shaped by the intersecting effects of culture, environment, gender, past experiences, or conditions and disabilities, such as fibromyalgia and autism (InTouch Digital 2020). The perspectives of both neuroscientists and haptic technologists largely rely on a literal or positivist rendition of touch as a physical skin-to-skin or hand-to-device form of contact. Yet the notion of touch plays a significant experiential and phenomenological role in more figurative, metaphorical and conceptual ways, and, in these terms, we suggest that mobile media do afford a partial (if at times insufficient) *possibility space* for mediated touch, a way of feeling together and being in affective proximity with each other, especially during a global pandemic.

Ironically, it is the complexity of our body schemas that also supports the notion that haptic and touch-screen technology can indeed "stand in" for touch, or rather, that the adaptability and plasticity of the body–technology relation can work to (at least partially) complete the felt sense. In phenomenological terms, such sensorial phantoms are possible because of our capacity for, or openness to, an "as-if" structure of experience, which is also enacted in our use of mobile media and haptic screens. That is, our ability to embrace ambiguous spatial perceptions and modes of embodiment within our corporeal schemata – the fact that we can oscillate between, conflate and adapt to disparate modes of being and perceiving – is precisely why telepresence is sensorially tolerable. Put another way, the very condition of mediated touch speaks to our capacity for sensory dispersion beyond the physical limits of the body, and our open-ness to embodied distraction. Throughout the history of media cultures, we have rapidly accommodated the "presencing" of radio, telephony, television, online, mobile and game technologies to such an extent that they are now quite ho-hum and habitual. As mobile touchscreens have become increasingly ubiquitous, we have similarly acquired a new repertoire of sensory awareness that expansively accommodates ambient proprioception and the telepresent intimacy of mediated social touch. As Powers and Parisi (2020) note, while we tend to think about touch as "a fundamentally biological sense, its meaning is continually renegotiated in response to shifting cultural conditions and new technologies."

The "stand in" for touch that mobile media offer may be perpetually incomplete, but the "as-if" structure of habitual experience rushes in to narrow the sensorial gap.

Consider how the closeness of shared and ambient play that many experienced both before and especially during the pandemic is achieved through a sense of networked co-presence. Such handheld intimacy is literally corporealized through chatting and playing *Scrabble GO* or *Roblox* with distant friends, as touch-screen gameplay with others in the network expresses not only a way of being in the world, but also a way of being-together *in the same microworld*. The type of intimate presence that we experience in networked gaming is echoed in a myriad of digital media practices: the seamless integration of "actual" and "virtual" experience as we engage in "meeting," "chatting," "playing," "being-in" places and "being-with" others in ways that do not differentiate between face-to-face and networked interaction, or between material and virtual forms of intimacy and social proprioception.

In her book *How to Feel: The Science and Meaning of Touch*, Subramanian evocatively describes her experience of the touch-enabled tablets created by Ed Colgate and his team at the Center for Robotics and Biosystems at Northwestern University, Illinois. The tablets work to simulate the feeling of touch, contour and texture in the fingers by changing the speed of up and down vibration, controlling the amount of air between skin and screen and thus altering the extent of friction that is felt (Subramanian 2021: 174). The devices can send and interpret "haptic pictures" that translate images into textural patterns, but Subramanian, who was keen to experiment with her partner, asked a computer scientist friend to make an app that would allow screen-through-screen touching. Initially she trialed the app with a colleague, with humorous results: she reflects that while the finger of another person touched through the screen doesn't "have the texture of a real one," it nevertheless feels like he's "reaching out" to her – a connection intimate enough that she quickly felt uncomfortable and "creeped out" (2021: 185). The experience with her partner Kartik was somewhat different. She writes:

> Our fingers meet up near the center of the screen, and we test it together. I can feel the little raised spot, like the glass screen is a little warped. We stay there, rubbing the impression of each other's presence. I let him take a few seconds to make sense of what he's feeling and then I move my finger away. He tries to come closer again. I dart away, and he chases me. I stop and let him catch me, letting his finger linger on top of mine. We gently nuzzle against each other. We are flirting, haptically! (Subramanian 2021: 185–6)

Reflecting on the experiment, Subramanian laments having to return the devices to Colgate's lab and misses them immediately, as the experience opened up a space of possibility for exploring more playful and intimate haptic connection with distant loved ones. She suggests that communicating through social media and text messaging often deprives us of the robustness of "true intimacy" and ponders if this is something that might (surprisingly) in some way be provided by "a tiny haptic vibration" (Subramanian 2021: 189).

Instances of meaningful and deeply "felt" mediated intimacy in everyday life are also frequently observed in ethnographic research that explores how mobile media users perform acts of care for vulnerable others, including children, isolated friends, the sick or elderly, and even companion animals. In their compassionate reflection on care and social connection during the pandemic – *Touch in the Time of Corona* – Steiner and Veel consider how our "possibilities for touch, touching, and being touched, both physically and affectively" (2021: 5) have been reconfigured by the many technological-material dependencies necessitated by the risk of contagion and the imperative of isolation and physical distance. Informal conduits of connecting and checking in via social media and mobile apps have become effective pathways of care-at-a-distance as we work to "preserve habitual forms of human interaction" and find alternative modes of experiencing emotional proximity (Steiner and Veel 2021: 9). We see this, for instance, in the example of the "digital hug." This involves social media users showing support to others in their network by posting messages accompanied by a #hug or #covidhug hashtag, or by including hug-related emojis, such as Facebook's care reaction emoji, where an animated "smiley clasps its small yellow arms tightly around a red heart." What is particularly noteworthy about the care reaction emoji in phenomenological terms is that Facebook has given "the smiley body features," and, in so doing, this animated emoji – enacted by the metonymic hand – "gestures towards a haptic experience that involves movement of the arms, and a body touching someone else" (Steiner and Veel 2021: 19). In terms of expressing care-at-a-distance, Steiner and Veel suggest that "in some cases, a digital hug carried out by technology seems to suffice, and sometimes even to exceed a physical hug, because it is so much more readily available" (27). Given the fraught nature of physical contact with other humans (i.e. physical touching in its verb form) during the pandemic, "as if" modes of touch have emerged, literally mobilized by the haptic affordances of hand-to-screen. Such forms of mediated intimacy and care-at-a-distance (i.e. touching in its adjectival or metaphorical form, as a non-physical form of connection) speak to the expanded "possibility of touching and being touched" (Barad 2012: 219).

As Puig de la Bellacasa puts it, touching through screens provides "a metonymic way to access the lived and fleshy character of involved care relations" (2017: 95). Similarly, sharing images of each other, our children, our pets and other companion animals on Twitter or Instagram becomes more than a way of interacting and sharing or maintaining social relationships – it provides a space to perform *feeling together*. Jan Sitvast notes how photographs and videos can effectively and therapeutically mediate social touch or the "felt sense," in the same way that we can be "moved" by music or poetry (Sitvast 2021: 2). A good illustration of this is the 2018 viral video of a cat gently lying down on an iPad after watching a video purporting to show its recently deceased owner. Even though this video was subsequently proven to be fabricated (Capron 2018), it provides an interesting and powerful example of how content can be manipulated to evoke affect and the felt sense. Indeed, for many of us, images of family or cute animals literally "resonate" in the body, bringing warmth, humor and affinity to the increasingly digitalized and detached contexts of the pandemic.

The pandemic also saw the further consolidation and growth in popularity of touchless payment systems and services. Early in the pandemic, touchless payments were touted in certain markets (such as Australia) as an effective means of purchasing goods and services, particularly within brick-and-mortar stores, while avoiding unnecessary human-to-human contact (such as, for example, through the exchange and handling of cash). Yet, even touchless payments retain certain "as-if" tactile qualities. Touchless payment systems employ near-field communications (Frith 2020), requiring the purchaser to place their credit card or mobile device against, or in close proximity to, the vendor's payment terminal. These human–machine interactions that occur within the communicative "near field" implicitly convey "touch potential" – or what Barad (2012: 208) calls "action-at-a-distance forces." Touch potential is a term derived from electrical engineering and refers to the possibility of the transfer of an electrical current when a human or animal subject stands next to or brushes against an object that has been struck by lightning. Here we employ this term in a specifically phenomenological sense, to draw out the idea that the mobile device or credit card not only acts as an extension of touch (like Merleau-Ponty's famous example of the sight-impaired man and his walking cane) but also reveals how, despite our best efforts to avoid it, touch perception and experience can be "proximal" or "proximate" – touch can be perpetually *in potentia*. Touch potential, as an affordance of mobile media, serves to remind us of both the complex nature of tactile relations and the often-overlooked agency of the machine in human–technology couplings.

Conclusion

As Richardson and Hjorth (2017) note, haptic technologies such as mobile touchscreens are transforming our sensorial experience of being-in-the-world, being-with-others and being-with-media. Many researchers are increasingly focusing on the intimate, social and playful nature of mediated touch. For some, the role of haptic technologies is especially important when it comes to vulnerable agencies, including the elderly and disabled. Touchscreens allow for non-normalized bodies to engage differently with games, affording multi-sensorial approaches that encourage different lived experiences and skillsets, thus enabling the inclusion of diverse users. Haptic screens also refocus our attunement away from the primacy of the visual and aural towards other modes of knowing and perception. In their critical assessment of the place of touch in the pandemic era, Powers and Parisi insightfully comment that the pandemic has brought about

> the most rapid upheaval in global practices of touching that we've seen in at least a generation, and it would be surprising not to see a corresponding adoption of technologies that could allow us to gain back some of the tactility, even from a distance, that the disease has caused us to give up. We owe it to ourselves now and in the future to be deliberate, realistic and hopeful about what touch and technology can do, and what they can't. (Powers and Parisi 2020)

In this chapter we set out to examine the complex relation between the hand and the mobile device, and the role and possibility of mobile haptics and the touchscreen particularly during the current decade, a time characterized by a profound lack of physical social touch and experiences of "skin hunger" as an effect of "touch starvation." We examined the significance of hands in the human sensorium, and the role of touch as a vital feature of human intimacy that has been highlighted since the start of the pandemic. We also sought to examine our capacity, via embodied and material metaphor and the phenomenological "dilation" of our body schema, to extend corporeal reach through the mobile interface, and considered the extent to which mobile devices – as enablers of ambient proprioception – might come to serve as a partial restorative for skin hunger.

Our exploration of the potential of mobile media use to "stand in" for touch has been in part a response to Powers and Parisi's call for both realistic and hopeful analyses of what technology can do. While it has been argued that touchscreen interactions are incapable of replacing "fleshy" social touch and are ineffective in assuaging skin hunger, mobile touchscreens nevertheless afford a sense of touch *in potentia*, generating a "possibility space" that transpires because of our perceptual capacity to engage in an "as-if" structure of experience via the metonymic function of the hand as a perceptual conduit for the whole body. Drawing on the phenomenological notions of mediated touch and social or ambient proprioception, we suggested that this experience of potential or "as-if" touch is importantly *non-trivial*, in that it provides us with a felt sense of being together (or being there for each other) when in isolation or not physically co-present. In this chapter we have explored just a few examples, though we acknowledge that the question and experience of touch relations varies widely across cultures and places. Despite (or because of) these variable contexts, there is rich opportunity for further, careful, phenomenologically informed examination of the perceptual role of mobile media in contemporary life and of how, why and to what extent touchscreen practices have the capacity to emulate – albeit in partial and fragmentary ways – touch-based interaction.

5
Feet

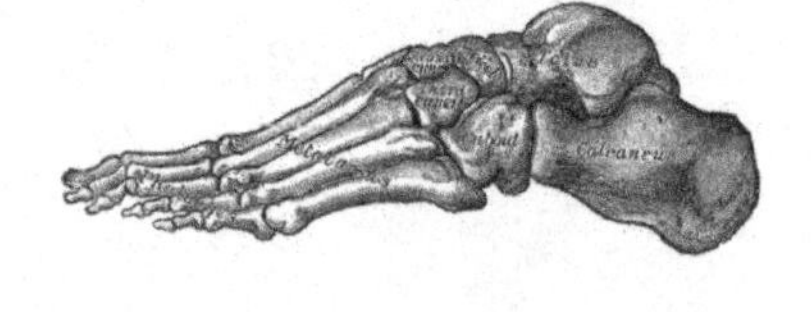

For it is surely through our feet, in contact with the ground (albeit mediated by footwear), that we are most fundamentally and continually "in touch" with our surroundings. (Ingold 2004: 330)

Human feet are remarkable structures. Each foot has twenty-six bones, thirty-three joints, and over 100 soft tissues, including two groups of muscles. The first group (extrinsic muscles) are connected to the foot but are found in the lower leg and serve to tilt the foot up (dorsiflex) and down (plantarflex) and roll it from side to side (inversion and eversion). The second group (smaller, intrinsic muscles) are contained within the foot and primarily serve to move the toes. Foot muscles are "key to how the foot functions during bipedal walking" (Farris et al. 2019: 1645). While "intrinsic muscles contribute minimally to supporting the arch of the foot during walking" – this function is served by extrinsic muscles and the plantar fascia ligament which forms the arch of the foot – they "influence our ability to produce forward propulsion from one stride to the next" (Farris et al. 2019: 1645).

There are also 7,000–8,000 nerve endings in each foot, making the foot extremely sensitive – and ticklish. This touch sensitivity is very important, especially when moving about barefoot, in that "mechanoreceptors in the skin provide sensory input for the central nervous system about foot placement and loading" (Henning and Sterzing 2009: 986). Stepping on a sharp stone or nail, for example, activates "protective sensation" in the foot, telling us to lighten our step or lift our foot. Thus, the human foot is an "intricate mechanism" (Chan and Rudins 1994) that serves as the primary point of contact between the body and the ground, cushioning the body and adapting it to uneven surfaces.

Our feet, of course, also permit us to walk upright, providing "traction for movement, awareness of joint and body position for balance, and leverage for propulsion" (Chan and Rudins 1994). Walking, simply put, works as follows:

> Muscles tense. One leg a pillar, holding the body upright between the earth and the sky. The other a pendulum, swinging from behind. Heel touches down. The whole weight of the body rolls forward on the ball of the foot. The big toe pushes off, and the delicately balanced weight of the body shifts again. The legs reverse position. It starts with a step and then another step and then another ... (Solnit 2000: 3)

While walking is not often thought of as a bodily "sense" in the same way that sight, sound, smell, taste and touch are, the human foot nevertheless serves as a vital "*sensory organ*" (Henning and Sterzing 2009: 986, our emphasis). Feet have strong tactile capacities and direct "movement adjustment and balance control." In this way, walking forms a fundamental part of our overall sensory apparatus.

For human bipeds – like other animals that use two legs for propulsion – walking is key to how we phenomenologically encounter and experience the world (DeSilva 2021). As theorized by human geographers, phenomenologists and proponents of material culture, walking is fundamental to our corporeality, ontology and cultural practices. This is reflected in the fact that walking holds a privileged place as a metaphor for capturing the biographical journeys or paths that our lives take. O'Neill and Roberts note that these metaphorical walks can be linear ("walking a (fine) line" with "travails along the way"), circular (where we have "come full circle"), up and down ("life on the up"; "life has been downhill"), "or some combination of such trajectories" (2020: 16). Lakoff and Johnson (1980: 14–21) refer to these collectively as orientational metaphors.

Walking is what Ihde (1990) would describe as a "multistable" phenomenon: it is something that we have in common (it is stable), yet we don't experience it in the same way (it is variable), differing between individuals and across cultures. For instance, hiking as a leisure-based form of walking, or touristic perambulations in a foreign city, are quite distinct from having to walk long distances for work (or to escape conflict), as they are from life in semi-nomadic tribal societies.

Walking in a city late at night is likely to be experienced very differently by a white male than it is by a woman or a person of color. As Hardley and Richardson (2021) note, citing Koskela's study of women's "spatial confidence" in the city, women will enact a "bold walk" that projects confidence and determination as a way to counteract fear, especially at night – a gendered bodily comportment that often involves the co-option of the mobile phone perpetually close at hand. Walking – and mobility more broadly (Goggin 2016; Parent 2016) – is also experienced differently by able-bodied people than by disabled or neurodivergent people (Stenning 2019).

Not only does walking engage all the senses, the human ability to balance and move about is also very much a multi-sensorial process. Human vision, for instance, functions "as an integral component of the control system for maintaining a stance" (Lee and Lishman 1975), and the inner ear (the vestibule and semicircular canals) is crucial for maintaining balance. More broadly, proprioception – our awareness of the position and movement of the parts of the body – is made possible by means of sensory organs (proprioceptors) in muscles and joints.

While, as noted, the "sensitive vestibular system of the inner ear governs balance," neuroscientists have discovered that "we have what are known as place cells in our hippocampi," and that "if you stay in one place, the cell for that position keeps firing, but if you move, that cell will stop firing and a cell marking your new position will start firing and so on" (Fleming 2019). At the same time, "our social brains are working to predict which direction others will take, to avoid collision." Thus, in order to walk and navigate successfully, the human brain "flickers between regions" (Fleming 2019).

Neuroscientists also believe that this has reciprocal benefits, with walking in turn stimulating other parts of the brain, including those responsible for creative problem-solving (O'Mara 2021) – whence comes the Roman saying, *solvitur ambulando* (the solution comes through walking), and why it is, presumably, that philosophers and writers historically have been so enamored of walking (see Solnit 2000: 14–29, for discussion). This may also partially explain the rise in popularity of walking meetings, and, especially during the Covid-19 pandemic, of walking video-call meetings (or "feetings").

The material, corporeal and cultural specificities of the foot as a significant component of the body-as-technical-instrument is explored by Ingold in his historical investigation, "Culture on the ground: The world perceived through the feet" (Ingold 2004). For Ingold, a critical inquiry into past and present "footwork" – as the opening quote of this chapter reveals – enables a "more literally grounded approach to perception" that counters both the privileging of vision in Western culture and the primacy given to manual touch in analyses of tactile perception, thereby restoring bipedal haptics to "its proper place in the balance of the senses" (2004: 330). Ingold's work aims to disclose the "special properties of pedestrian touch" and "circum-ambulatory knowing" as a way of countering the higher corporeal and technical status given to the hand – a bias that is clearly evident in phenomenologies of the body–technology relation and current investigations into the "handiness" of mobile phone use. He writes:

> Rather than supposing that the hand operates on nature while the feet move in it, I would prefer to say that

> both hands and feet, augmented by tools, gloves and footwear, mediate a historical engagement of the human organism, in its entirety, with the world around it … It is in the very "tuning" of movement in response to the ever-changing conditions of an unfolding task that the skill of walking, as that of any other bodily technique, ultimately resides. (Ingold 2004: 332)

Ingold's work poses important questions for understanding our embodied engagement with the world around us: how are our bipedal *habitus du corps* and footwork techniques implicated in our material and spatial perception of the environment, and in our ways of knowing and being more broadly? More specifically, what happens when *mobile phones* become part of the experience? What does the mobile phone modify in terms of walking, and what changes follow from these modifications? When pedestrian mobility becomes mediated by a device, we can observe how one's experience of walking on a surface becomes modified; our gaze becomes split between the mobile screen and the environmental affordances and terrain features (paths, obstacles, steps, slopes) that we negotiate as we walk (Gibson 1986: 36–7).

To cast the above questions in phenomenological terms, we might ask: what is "dilated" (extended) and what is "contracted" (reduced, closed down) through our use of mobiles while walking (keeping in mind that mobiles permit these two things to occur simultaneously)? In the early literature on mobile phone use, for instance, it is argued extensively that mobile phones dilate our experience of place and expand our sense of social connectedness. Phones construct "complex modulation[s] of presence" (Lasen 2006:

227) by facilitating co-present encounters and conversations with proximate and distant others (Licoppe 2004; Gergen 2002), thus leading to "new senses of place being constructed as a hybrid between co-located and remote social contact" (Ito 2003).

In this chapter, we examine a number of these dilations and contractions in our discussion of how our use of mobile devices serves to (re)shape our experience of walking. We explore how our use of mobile devices affects pedestrian mobility, contracting one's sense of one's immediate surroundings, and how mobile maps, geolocative social media applications and location-based games have the potential to dilate our sense of place, altering our journey trajectories, our navigational and coordination practices, and, potentially, our co-present encounters. We also detail how augmented reality (AR) applications and other forms of automated data capture dilate our sense of place and pedestrian experience by expanding the information associated with particular places. And we close by examining how a politics of mediated walking works to both dilate and contract pedestrian experience, by opening up opportunities for some while closing them down to others.

The Mobile Phone Pedestrian: Everyday Distraction and Co-present Interaction

A central contention of this book is that the dynamic shaping of our corporeal schemas is under continuous modification by artifacts, tools, techniques and more complicated technological ensembles, which are always-already embedded in a palimpsest of cultural milieus and collective habits. As we have argued throughout this book, mobile phones involve their own specific sensory modes of embodied use, demanding of us that we pay attention to the particularities of body–technology relations. This understanding is particularly useful for interpreting both the somatic intimacy of handheld media and the collective or sedimented routines of walking and urban pedestrian movement when modified by mobile media user practices.

Here we aim to draw attention to a "peripatetic phenomenology" of walking through an analysis of the relation between pedestrian mobility and the use of mobile devices, illustrated through a range of established and contemporary examples. Our contention is that mobile media use has a significant bearing on the habitudes of walking as such devices become increasingly embedded in our everyday ambulatory activities.

One of the characteristics of pedestrian mobile phone use is that the specific and complex nature of user engagement with mobile handheld screen devices and/in urban places is shaped by everyday distraction and co-present interaction (interaction with others, who may or may not be in the same physical space). Morse's (1998) conception of the "copresence of multiple worlds in different modes" experienced as an "ontology of everyday distraction" is particularly helpful in this context for making sense of contemporary screen-based mobile technologies like smartphones. The first part of Morse's formulation draws attention to how our interactions with space and place are heavily mediated by technology in ways that permit us to be in two (or more) places at once. Rowan witnessed this first hand while passing through Melbourne on foot recently. A woman walking briskly in front of him was weaving down the street, deftly negotiating street furniture and other pedestrian traffic. As she strode along, she was playing music out loud through her phone, nodding her head to the beat and fist-pumping the air with every cymbal crash. She happily straddled two places at once. An ontology of everyday distraction permits us to occupy multiple worlds in different modes, through the intermingling of physical mobility, face-to-face encounter, and telephone- and app-mediated voice, sound, text and image transfer. Such complexities of interaction are illustrated in Ito's ground-breaking research on Japanese youth and their mobile use (Ito 2005). This research revealed "engagements with mobiles that involve a complex set of interactions within and between places, incorporating physical transportation (on foot and by other means), navigation (utilizing, among other tools, existing landmarks), fluid temporal arrangements, and communication (both mediated and face to face)" (Wilken 2014: 517; see also Ling and Haddon 2003).

Equally relevant to understanding pedestrian mobile phone use is the second part of Morse's formulation: the notion of an "ontology of everyday distraction." This is the idea that a precondition of our engagement with technology, space and place is an experience of distraction and disjointedness. That is to say that our technological tools and our surroundings are influenced strongly by a variety of environmental, contextual, interpersonal and cognitive factors (Bassett 2005). Thus, for technologically equipped pedestrians, attention is always a fluctuating state; they tend to switch their attention from one place or activity to another, each time at the expense of the last, and as a result of competing stimuli. On the one hand, this can mean that mobile users pay more attention to "phone space," which is "often prioritized over local space" (Bassett

2005). For example, some years ago, while attending an overseas conference in an unfamiliar city, Rowan misunderstood the venue information presented on his mobile map and consequently took himself and a colleague in the opposite direction to where they had intended to go, ending up in a completely different (and rather less than savory) part of the city. Only then did he realize that there had been a misrecognition of venue and location. On the other hand, however, "since attention never presumes absolute presence, it cannot presume absolute disconnection" (Bassett 2005); rather, there is a toggling between the two. Thus, pedestrian use of mobile media elicits variable levels of attention and inattention that shift between actual and telepresent space, partially depending on the demands of the immediate environment and the extent to which the interface becomes ready-to-hand in a Heideggerian sense (that is, its function and usability become fully embodied and intuitive and recede from explicit awareness).

Complementing Morse's conception of the "copresence of multiple worlds in different modes" experienced as an "ontology of everyday distraction" is Cooley's notion of "screenic seeing," which forms a useful model for making sense of our visual engagement with mobile screens. "Screenic seeing" captures the idea that vision is "no longer a property of the window and its frame" (Cooley 2004: 143), but, rather, contributes to a more expansive "material experience of vision" where "hands, eyes, screen, and surroundings interact and blend in syncopated fashion" (145). Cooley marries this with Benjamin's notion of tactility, where the tactile is associated with distraction and habit, to argue that "screenic seeing" is "never focused." Rather, it "spreads out, across various images (of landscape and screen), as a result of and in tandem with the interpenetration of the hand and the mobile screenic device" (147).

The distracted nature of "screenic seeing" has implications for understanding pedestrian mobility, with clear evidence that "traditional" modes of pedestrian navigation are disrupted by mobile media use. Perception of surface and texture, as felt through the feet and passed over by the eye, are crucial to all successful pedestrian movement and navigation. An ever-shifting gaze, which moves over surface and texture, is critical to how we perceive visual depth (Morris 2004) and distance (Higuchi 1983). It is also crucial to how we negotiate fellow pedestrians in urban space. Goffman's (1963, 1972) and Whyte's (1988) pioneering observational studies of pedestrian interaction found that walking in crowded city streets or urban plazas involved a "complex web of movements" (Whyte 1988: 59), where the smallest facial or other bodily gestures are crucial to successful navigation and avoiding collision.

Goffman's and Whyte's detailed attention to the micro-gestures and eye-behavior necessary for pedestrian navigation can also be insightfully applied to a peripatetic phenomenology of mobile phone use in public urban places. It has been noted within the established literature on mobile telephony that the "downcast eyes" of users attending to their mobile screenic devices disrupts the "scanning gaze" (Höflich 2005) that is considered necessary for successful pedestrian flow (Tiessen 2007). In other words, mobile phone use unsettles established patterns of public interaction and navigation, thereby causing fellow walkers to renegotiate their spatial and pedestrian movement by taking into consideration the lack of familiar facial cues. Attending in this way to the pedestrian and peripatetic demands of mobile media offers broader insights into our micro-mobilities and "techniques of the body" enacted in public urban and pedestrian space. This occurs both in terms of our smaller gestures, motor movements and techniques and footwork particular to mobile phone use, and our larger trajectories through the city or full-bodied actions such as walking while talking, texting, gaming or consuming streamed content.

In his study of human movement within the Piazza Giacomo Matteotti, Udine, Italy, Höflich observes that the mobile phone pedestrian functions as both a "vehicular unit" (as they move through the piazza) and a "participation unit" (as they interact with and navigate around others already occupying the piazza) (2005: 164). Two key forms of pedestrian engagement were observed. First, that of mobile phone pedestrians who transected the piazza on their way to somewhere else, immersed in phone conversation while subtly navigating their way across the crowded space, keeping "an eye on where people around them are heading," constantly maintaining a visual "scan-and-control-area range" ahead, and adjusting their movements accordingly (164). Similar findings were reported by Lasen (2006) in her three-city study of mobile phone use in public space, where she noted that it was common across the three sites (London, Madrid and Paris) to observe people "looking alternately at their phones and their surroundings." Second, Höflich observed those who lingered in the piazza while on their phones. These mobile phone pedestrians tended to gravitate towards central orientation points, move in circles or S-patterns to avoid collision with others, or walk to-and-fro (2005:

167). Moreover, as others have noted, mobile phone pedestrians in such situations also tend to slow their walking, stare at the ground or into space, and "adopt a position of physical withdrawal" (Kopomaa 2000: 81) to create a "protective shield" (Höflich 2005: 167) and a private place for conversation within a crowded public place. One's own body may thus "behave" in ways that accord with (or deviate from) consensual and recognized modes of being-on-the-phone. Such "repertoires of gestures" (Ling 2002: 64) include, on the one hand, stopping, bowing the head, shielding one's mouth or face with the hand to define a provisional private space, and, on the other, deliberately not altering one's trajectory or visual orientation. Across both forms of pedestrian engagement, interactional possibilities are dilated *and* contracted. What is more, the various postures and embodied actions, and the dynamics of attention–inattention, are quite specific to the body–mobile relation which has emerged over the last decade or so. These repertoires of gestures are not enacted in universal ways, however, and remain very much multistable phenomena, with subtle yet distinct differences across cultures.

In a very real sense, mobile-mediated walking changes our perception of terrain features: with the downcast eyes and distracted gaze of screenic seeing there is a greater emphasis on the "special properties of pedestrian touch" (Ingold 2004: 330). We can observe this, for instance, in how pedestrian mobile users, while in conversation, will tend to keep moving in order to "lessen the intensity of visual perception" (Kopomaa 2000: 82). This could also be taken as an indication that pedestrian mobility is an inherently adaptive practice and that mobile technology use contributes to shifts between modes of perception, as well as further adaptations and alterations to perceptual processes. As de Certeau (1984) famously suggested, a pedestrian *makes possible* the space of the city in collusion with the built environment. The steps one takes – the act of walking through the city – are altered by mobile phones and other portable devices. De Certeau argued that "a spatial order organizes an ensemble of possibilities" (1984: 98) – the places where one can go and the objects blocking or redirecting one's path. At the same time, the pedestrian trajectory actualizes and creates some of these possibilities simply through the "improvisation of walking." Yet, potentially, this trajectory can be quite radically revised and re-possibilized, as we discussed in the earlier chapter on ears, by the interruption of a mobile phone call or the ping of a text message, by listening to and downloading music, by changes in the immediate soundscape, or by those "telepresent" on the other end of the phone becoming "virtually" integrated into one's route. We see this, for example, in the contemporary phenomenon of the "Bluetooth pedestrian" (Richardson 2005), who is able to take calls while on the move, seamlessly integrating walking and talking. However, when a Bluetooth pedestrian speaks (or gesticulates while speaking), it can be disconcerting for nearby pedestrians as they cannot hear what is occurring through the other's wireless headphones or AirPods, making it appear as if the Bluetooth pedestrian is talking to themselves. The lack of obvious auditory cues signaling that a two-way conversation is occurring can be – if only momentarily – potentially disrupting to ordinary pedestrian navigational processes, requiring heightened situational awareness. In phenomenological terms, what is being described here with the Bluetooth pedestrian is the body's capacity to intertwine with the world, to integrate, internalize or *intercorporealize* seemingly external objects, spaces and bodies and environments into our corporeal activities – as well as the capacity of bodies to respond to the intercorporealization of others. Here, intercorporeality captures the coupling of tools and bodies, and describes the irreducible relation between technologies, embodiment, knowledge and perception.

Finding Our Way

The process of intercorporeality sits at the heart of empirical accounts of how people encounter and use mobile phones and smartphone map applications. As noted earlier in this chapter, in the pre-smartphone era of mobile telephony, the establishment of basic daily travel and social arrangements was shaped by processes of "micro-coordination" (Ling and Haddon 2003). Prior to cell phones, as Ito and Okabe observe, "landmarks and times were the points that coordinated action and convergence in urban space. People would decide on a particular place and time to meet, and converge at that time and place" (2005: 266). With the arrival of mobile phones, however, it becomes more likely that an initial and rather loose arrangement is agreed upon, and then, as the time of meeting nears, a "coordinated dance" occurs, with "contact via messaging and voice becom[ing] more concentrated, eventually culminating in face-to-face contact" (Ito 2003: 267). It's also common for mobile communication to continue even after physical co-presence has been achieved, particularly in order to enroll "distant others" in "socially co-present" encounters (Ito 2003; see also, Lasen 2006).

With the arrival of mobile map apps, "everyday urban navigational practices" are found to be "emergent and context dependent encounters" (Duggan 2018). Following Ingold (2007, 2011), Duggan (2017) argues that contemporary mobile map use is a dynamic, contextual and "processual" activity, with movement "a key factor in mediating practices of navigation" (Duggan 2018). Maps (mobile and otherwise), Duggan found, were consulted at key "decision points," in order to test one's hunches and assumptions, or (re)confirm one's position. In this sense, navigation itself, it would seem, is "solved by walking" (*solvitur ambulando*). Rowan remembers walking in Tokyo several years ago and having to regularly consult prominent landmarks as the canyoning effect of surrounding buildings interrupted the GPS signal, making the mobile map an unreliable navigational guide.

The assumption is that the blue dot on map apps, or the "actionmoji" of Snap Maps (Wilken and Humphreys 2021), work to construct a sense of "ground truth," fixing our position in time and space. When a stranger asks for directions on a street, for example, it has become common practice to consult a mobile map. However, an "embodied location of oneself in space" (Wilmott et al. 2018: 89) has been shown by mobile maps researchers to function very differently: it is not only processual, dynamic and contextual, but also highly individualized (idiosyncratic, even), and with some resistance to the body–tool coupling. This takes several configurations: there are phone users who rely on mobile maps (Duggan 2018; Özkul 2015b); those who believe in the importance of experiential, embodied knowledge over maps, with the suggestion that "people should know where they are" (Wilmott 2020: 119);

those who embrace urban exploration (Wilmott 2020; Özkul 2015b); those who employ some combination of the preceding approaches – what Wilmott refers to as "navigating by memory, maps and imagination" (2020: 95); those who pre-plan their journeys before leaving home (Duggan 2018; Wilmott 2020: 155); those who don't trust their own sense of direction yet also "don't trust that little blue dot" (Wilmott 2020: 87); and those who are frustrated with the "toponymic quagmire" (232) of place names offered by mobile maps and the lack of incorporation of embodied, experiential knowledge within mobile mapping applications.

Intercorporeality has also been negotiated in complicated ways through embodied engagement with mobile, location-based social networks (LBSN), such as the once-popular Foursquare. One way that this occurs is through the capacity of LBSN to foster urban exploration and place discovery (Frith 2014) by promoting "alternative ways of experiencing the city" (Özkul 2015a: 113). While locative media platforms actively cater to, encourage and benefit from these exploratory desires, end-user practices nonetheless have been revealed to be significant in generating "a new sense of place [and] a new way of place-ing oneself" (Özkul 2015a: 107) in an embodied sense. Mobile devices also foster new forms of spatial knowledge through "digital wayfaring" (Hjorth and Pink 2014: 45), whereby "mobile media users perpetually move in, around and through the environment, each making their own physical and digital trace," such as by taking photographs or by "checking-in" to a venue, thereby establishing new forms of "online place attachment" (Schwartz 2015). In addition, LBSN applications fulfil vital mnemonic functions for their users. LBSN like Foursquare (in its original incarnation) have been understood by their users to function as a persistent and searchable digital log of sorts for documenting and recalling places visited (Frith and Kalin 2016; Özkul and Humphreys 2015), and for creating a "curated" history of "personal movements" and personal moments (Evans and Saker 2017: 50). In addition, LBSN applications, and camera phone practices more broadly, provide key means for recording "ordinary affects" (Stewart 2007) and the rich "tapestry of everyday life" (Hjorth and Pink 2014: 53) while on the move.

Playful Walking

Location-sensitive mobile games evidence an "enfolding of contexts" (de Souza e Silva 2004: 18) whereby the game space that is accessed through the mobile device overlaps and intertwines with urban space in important ways. Location-based mobile games also encourage, and in many cases demand, quite specific forms of bodily engagement that require players to actively walk in geographical space to control their in-game movements; these involve the skillful and rapid navigational oscillation between the mobile screen and the surrounding urban environment.

This complicated, two-way interaction and experiential flow between game space and urban space also has a direct impact on the way these games are played. In work on the early location-based game *Mogi*, researchers found that players made significant adjustments to their daily temporal and spatial routines and rhythms in order to maximize the opportunities presented by the game (Licoppe and Inada 2006:

57). This included players changing their established mobility patterns and preferred routes through urban space in order to prolong the game experience and maximize their opportunities by, for instance, taking a detour in order to collect specific and sometimes hard to acquire or rare virtual objects. In this sense, the game takes on some of the characteristics associated with the Situationist practice of *dérive*, with Kim suggesting that *Mogi* amplified people's "ordinary behaviour – it changes going on an errand into a piece of a game" (Kim, cited in Hall 2004).

Location-based games often rely on a treasure-hunt dynamic, which has a long history as a mode of hide-and-seek discovery on foot. The game *Geocaching*, currently played in over 200 countries, requires players to hide tradable items in geocache containers marked with GPS data in publicly accessible places, for others to track, find and log using their mobile devices. In his analysis of *Geocaching* and other location-based games, Farman (2012) suggests that a shared and collaborative augmented reality emerges, where walking bodies, digital networks and material space converge.

Of all mobile location-based games, *Pokémon GO* has been by far the most popular, with the game still being played by around 9 million people daily (averaging 78 million unique users per month) and generating over $700 million in revenue per year. Players download the app onto their iOS and Android devices and wander their local neighborhoods and public spaces in search of Pokémon and PokéStops, encountering the eponymous monsters in real-world settings, and competing and connecting with other players through virtual Pokémon gyms, raids and battles. *Pokémon GO* play involves a form of pedestrian labor that interweaves digital and physical information, quite literally altering the "spatial legibility" of urban space – or the way urban environments appear as coherent and recognizable patterns (Frith 2014). Like *Mogi*, the game encourages players to take routes through streets they would not otherwise visit, and to think of their local parks, shops and neighborhoods as repurposed locations for play, transforming banal and familiar surroundings into game terrain – a gym or PokéStop might be situated at the local library, café or nearby graveyard. Such experiences also point to a quite extraordinary effect: as some media theorists have suggested, there is evidence that *Pokémon GO* acts as a catalyst for large-scale changes in people's "destination choice" or "trip distribution" (Colley et al. 2017). In other words, such games have the potential to motivate "people to do something they rarely do: substantially change where they choose to go" (Colley et al. 2017). They invite a form of walking-as-play – playful pedestrian behavior that nevertheless comes with some physical risk, as has been documented rather sensationally in popular media, with stories of *Pokémon GO* players stumbling into oncoming traffic or over cliff edges in their quest for game resources. Such risks speak to the significant disjuncture between actual and augmented reality (or the imperfect translation of physical space onto the small screen), but are also indicative of a shift in the eyes–feet relation that is demanded by mobile devices more generally, and the corporeal "trouble" that can ensue.

More recently, game designer Naomi Alderman's creation *Zombies, Run!* has successfully integrated audio storytelling into location-based mobile gameplay to augment the experience of jogging in the urban environment. In developing the mixed-reality mobile app, Alderman (2019) effectively exploited both the propensity of joggers to listen to playlists and podcasts on their mobile device, and the popularity of zombie narratives in the contemporary imagination, in particular our familiarity with "heart-pounding running sequences in zombie movies." Progression in the game is cumulatively achieved the more the player runs and collects items in their effort to escape the undead and successfully complete the mission directive. As Alderman comments, it "feels like the characters are whispering directly into your ear." *Zombies, Run!* maximizes the perceptual and affective intimacy of mobile media as a synesthetic experience involving eyes, ears and the moving body.

The preceding discussion reveals how location-based games are thoroughly embedded in the dynamic, continual and relational process of place-making. As Massey notes, a relational understanding of place involves "articulated moments" that include "networks of social relations" (1994: 154). This is significant for considerations of multiplayer location-based games, like *Mogi* and *Pokémon GO*, insofar as relational place-based interactions extend beyond the experiences of the self to include encounters with people in close geographical proximity in a form of walking-and-playing with others. Indeed, for avid players of *Pokémon GO*, restrictions on movement as a result of the Covid-19 pandemic impacted opportunities for social interaction – "socialization was recontextualized" – and changes to the game to accommodate stay-in-place restrictions were "generally received poorly by existing players, who did not appear to receive the same sense of gratification from the online platform as they did from in-person interactions" (Dunham et al. 2022: 21, 11).

Playfulness can also be experienced in other ways, such as by exploiting the ludic potential of apps intended for other purposes. It has become popular, for example, to use fitness tracking apps like Strava to create map-overlay images, known as "Strava Art." The idea is to walk or ride along predetermined routes within a city to generate self-tracking data that, once all routes are completed, forms giant line-drawing images when viewed on a map of the city. The resultant images appear to sit atop the landscape, like a contemporary data-driven equivalent of the Cerne Abbas Giant; for example, a bull in central London whose rear hoof rests on the northern boundary of Hyde Park while one of its horns is located in Camden Town; a big bunny in South London; and an enormous woolly mammoth that spans most of Yunlin County and Chiayi County in south-western Taiwan (see strav.art). In this way, Strava users incorporate two simultaneous perspectives – a "grounded" view of their own bodily positioning in urban space, and a bird's-eye view of their position and movement on a map. Both perspectives, as with location-based gaming, are achieved by adeptly shifting between screen and environment, employing dynamic modes of "haptic vision" that involve eyes, hands, feet and mobile screen devices.

Walking and (Dis)embodied Sensing

It is the corporeal schema's dilatory capacity that effectively renders many of our technological interfaces transparent. As noted in Chapter 3, for example, once we become accustomed to wearing glasses or a wristwatch, these external objects tend to recede from awareness and become part of how we engage with the world around us (Adams 2019: 119). With wearable technologies, such as the Fitbit or Apple Watch – modern incarnations of the humble wristwatch that have been dubbed "everywear" due to their close tethering to our bodies and habitualized use (Gilmore 2016: 2525) – the process of intercorporealization appears to be a little more complicated. On the one hand, the Apple Watch "shapes the way we locate ourselves: from eye to finger, from heart to skin, from ears to arm" (Wilmott et al. 2018: 89); it does this because it "rests against our skin, monitoring our heartbeats, tracing our fingertips, counting our footsteps and tracking our location" (80). On the other hand, the Apple Watch's constant demands, in the form of a repetition of inputs and notifications, have been described as requiring "a labour of attention" from the wearer (Wilmott et al. 2018: 89; Gilmore 2016: 2526). Ihde characterizes this call for interaction as a shift from *embodiment relations* to *hermeneutic relations*: "we have moved from experiencing through machines to experiencing machines" (Ihde 1979: 11).

For critics of these devices, wearables draw renewed attention to the human body, yet do so in ways that diminish the phenomenological richness of our encounters with the world and in fact lead to bodily disappearance. This is said to occur in at least three ways.

First, step-counting wearables such as the Fitbit, it is asserted, flatten the complexities of our pedestrian *habitus du corps* and the footwork techniques that enable us to "know as we go" (Ingold 2010: S133), reducing them to steps counted or yet to be counted (Gilmore 2016: 2533; Adams 2019: 118). Adams draws on the work of Leder (1990) to argue that they make our bodies "*dys*-appear" – a term Leder coins to "describe the process through which a painful or ill body becomes present" (Adams 2019: 120). Adams' argument is that step-counting (and its close relationship with health-related risk discourses) reproduces, on an intimate, individual scale, similar interruptions in "mind/body/world harmony" that are said to be characteristic of a body in pain (120).

Second, wearables create a sense of bodily disappearance through their emphasis on aggregate data. Wearables ask us to rethink and share our bodies as data (Gilmore 2016: 2526). As Gilmore puts it, rather than dwell on the embodied experience of "aching legs," we pore over the "numbers collected by pedometers, accelerometers, gyroscopes, and geolocative devices." Everywear "permits individual access to knowledge about the body" (Gilmore 2016: 2534) through the data each device generates – knowledge that fueled the rise of self-tracking (Neff and Nafus 2016) and the quantified self movement (Lupton 2016) – while, in reality, this knowledge is limited. As Crawford et al. note, companies get to see the aggregated data generated by these devices, while users of wearables receive little to no information about this data or about "the cultural and scientific assumptions that undergird notions of normal users" (2015: 494). As Suneel Jethani (2021: 59) puts it, data aggregation erases difference. The collection, collation and aggregation of data from wearables flattens the specificities and complexities of human–technology relations, bringing about a sense of bodily disappearance.

A third form of bodily disappearance involves disabled bodies. This form of disappearance occurs through the lack of accommodation of the specific needs of those with disabilities by the makers of many wearables. A

2018 article, for instance, noted that "of 1000 different activity trackers [on the market at the time], only one, Apple Watch, can track the specific movements of wheelchair use and convert it into calories burned" (Passanante Elman 2018: 3762). Bodily disappearance also occurs through an orientation towards what Passanante Elman terms a "politics of compulsory able-bodiedness" and an associated tendency in the commercial discourse on wearables to "sell a vision of perpetual rehabilitation using the individuating neoliberal rhetoric of 'personal responsibility'" that equates "(nondisabled) mobility with health and freedom" (3774).

Conclusion

In this chapter we have explored various ways in which our everyday pedestrian activities shape our use of mobile phones and vice versa. We have detailed how our embodied, ambulatory engagements with mobile devices contribute to certain "dilations" (extensions) and "contractions" (reductions) in how we experience our surroundings and each other. We investigated how mobile devices affect pedestrian mobility, contracting one's sense of the immediate surroundings and causing certain interruptions of and reconfigurations to our micro-mobility patterns and navigational interactions with passers-by. In addition, we explored how mobile maps, mobile location-based social networks (LBSN) and location-based games carry the potential to dilate one's sense of place, altering journey trajectories, navigational and coordination practices, and, potentially, our co-present encounters. We also detailed how augmented reality applications and other forms of automated data capture dilate our sense of place and pedestrian experience by expanding the information associated with particular places. And we touched on how a politics of mediated walking works to both dilate and contract pedestrian experience, opening up opportunities for some (such as white men) while closing them down for others (such as women and people of color).

From this peripatetic phenomenology of the relation between pedestrian mobility and mobile devices, it becomes clearer just how extensively mobile media use has a significant bearing on the habitudes of walking as such devices become increasingly embedded in our everyday ambulatory activities.

In the following, final chapter, we draw the various strands of this book together, and begin to consider how our respective senses and sensory modes of engagement operate as an ensemble. While we have looked at individual body parts and specific senses in this and preceding chapters, we understand that mobile media use requires an intermingling of the senses. Indeed, one larger, unifying argument of this book – and the specific focus of the next chapter – is that our use of mobile devices involves an affective and embodied engagement with sight, sound, touch (via hand and foot), movement and cognition. As part of this, we give some consideration to what forms embodied, multi-sensory engagement might take for those experiencing neurodiversity, like synesthesia, and we close by drawing on recent work on multispecies sensory engagements with mobile devices as one possible line of future enquiry.

Conclusion

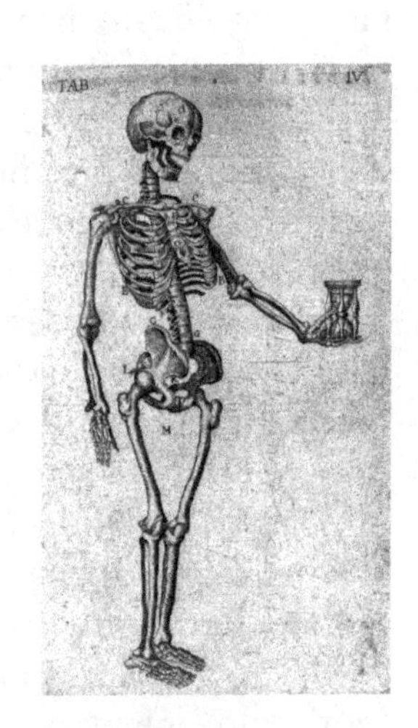

One characteristic feature of the telephone: it requires the participation of all the senses and mental faculties. (Kopomaa 2000: 83)

In this book, we have explored our embodied, sensory interactions with media technologies, with a particular focus on mobile interfaces. In a 1967 interview, media theorist Marshall McLuhan commented that "a medium is not something neutral – it does something to people. It takes hold of them. It rubs them off, it massages them, it bumps them around" (cited in Rosenthal 1969: 20).

Conclusion

In part inspired by McLuhan's ground-breaking work, this book has examined the human–technology relation specific to bodies and mobile media. In it, we have considered how mobile media use expresses not only a way of being in the world, but also a way of being together that requires ongoing corporeal adjustment and bodily incorporation. Mobile media and network practices become enfolded inside present contexts and activities, such as the embodied and itinerant acts of walking, and numerous other material and bodily involvements. Following Ihde, in each chapter we have detailed how our contemporary engagement with mobile media is a particular kind of human–technology relation that is both stabilized and destabilized by individual and collective biocultural variation. We have explored such variability in terms of perceptual and experiential differences that depend on contexts of use, access and bodily specificity, and described how modes of embodiment and perception are not uniform or neutral but saturated with conceptual and perceptual histories, individual variation, collective habitudes and sedimented ways of being in the world and being with others.

As we have aimed to convey in the anatomical and physiological descriptions opening each chapter, sense perception is incredibly rich and complex. Our understanding of the cultural implications of this perceptual spectrum is no doubt enhanced by even a basic comprehension of biology. Our focus on individual modalities of perception, as articulated in the book's introduction, is a strategic approach enabling us to "drill down" into the nuances of each sense at a micro-perceptual level. Yet the human body is of course a complex, sensorily *integrated* system. Every waking moment of each day we receive, organize, process and respond to information received from all our senses. Sherrington notes that "all parts of the nervous system are connected together and no part of it is probably ever capable of reaction without affecting and being affected by various other parts, and it is a system certainly never absolutely at rest" (1920: 8). Welch and Warren point out that "one is rarely if ever confronted in the everyday world with only a single form of sensory stimulation." Rather, our interactions with the world around us routinely involve a mingling of the senses: "a person might be looking at an object while simultaneously feeling it and hearing someone's voice off to one side. In this instance then the visual and haptic senses are both conveying information about the shape of the object, and the auditory modality happens to be stimulated at the same time" (Welch and Warren 1986: 25–2). This is to say that, while our senses are distinct, the fact that "they are all usually deployed all the time" means that "no sense ever senses in isolation" (Ekdahl 2015: 26). The thoroughly routinized nature of our intersensory interactions seeps into the everyday metaphors we use. The English language, for example, contains manifold examples of cross-modal sensory metaphors, like sharp taste, soft sounds, sour smell, sweet music, and so on (Cacciari 2008: 427–8). We have discussed the significance of metaphor at various junctures in each chapter, and note here again, following Lakoff and Johnson (1980), that metaphor in all its forms is grounded in our embodiment and works to structure our most basic understanding of the world.

In this final chapter, we seek to weave together the sensory strands of mediated experience traced throughout the book. That is, we turn to a more holistic interpretation of mobile media practices as multi-sensory and *synesthetic* – moving beyond an account of specific sensory ratios or body–technology couplings to begin to consider how our respective senses and sensory modes of engagement operate as an ensemble. As part of this, we give some consideration to what forms embodied, multi-sensory engagement might take for those experiencing neurodiversity, like synesthesia. In exploring sensory ensembles, we are also interested in how our sensory modalities operate in combination to create more fully embodied technology relations and engagements with mobile devices. Here, we address those sensory modes of engagement not given specific chapter-level attention – namely, taste and the olfactory sense – using these as a point-of-entry for thinking about emerging work on corporeality and body–technology couplings, especially as they relate to embodied, multispecies engagement with screen media.

"Sensorium Commune": Synesthesia as an Entanglement of the Senses

Every morning as Felix is about to step out the door, he inserts wireless earbuds into each ear and cues up a Spotify playlist to suit (or shape) the mood of the moment. While on the move, he likes to catch up on and respond to text messages that have been sent from friends late the night before (this works because his mobile plan privileges talk and text; limited data means he prefers to read and post to his Snapchat feed when there's Wi-Fi access). When walking, his head is lowered towards his phone, which is cradled in both hands. Even so, he tends to type by feel (and muscle memory) as much as by sight. This is because, while his head is lowered, his eyes are frequently raised, scanning the path ahead and reading the micro-gestures and bodily cues of those walking towards him so as to avoid collision. Between texts, the vibration of phone in hand or pocket alerts him to the arrival of new messages. His scanning gaze is also always alert to photographic or mobile video opportunities, the chance to capture unusual sights, such as someone in a peculiar outfit, and other oddities of the street, like a dog being "walked" in a pram by its owners. So adept has Felix become at capturing these fleeting moments that he is able to activate the phone camera with little break in his stride and to reposition the device so subtly and unobtrusively that he rarely stirs the camera consciousness of his unsuspecting subjects.

Embodied mobile use is rarely if ever experienced through a solitary sense. Rather, as described in the vignette above and noted at the outset of this chapter, mobile phones require the "participation of all the senses and mental faculties" (Kopomaa 2000: 83). While for the sake of clarity and ease of examination we have separated out primary modes of somatic engagement, privileging senses (or body parts) in each chapter, it remains the case that sense perception is always-already a whole-of-body experience because "the senses communicate with each other" (Merleau-Ponty 2004 [1962]: 262). We exist as a "*sensorium commune*" (277), a communion of senses, and our use and experience of mobile media is very much multi-sensory and *synesthetic*.

For neurodiverse people, the "conjugation" of the senses can be experienced via a more specific understanding of synesthesia, one that seeks to capture the phenomenon of encountering one sense through another, or what is known as "cross-modal association" (Cytowic 1995a). When "the involuntary physical experience of cross-modal association" occurs, a synesthete might experience letters, numbers or sounds as colors, or, to cite a well-publicized example, they might taste shapes (Cytowic 1995b). There are many different types (and subtypes) of synesthesia understood as cross-modal association (Barnett et al. 2008), including grapheme-color synesthesia (where individual letters or numbers are understood as specific colors), chromesthesia (the experience of sounds as colors), and auditory-tactile synesthesia (where certain sounds can induce sensations in other parts of the body).

Synesthesia can develop during childhood, but it can also sometimes be triggered in later life through the taking of psychoactive substances. In *Phenomenology of Perception*, Merleau-Ponty notes the synesthesia that results from consumption of mescaline (a psychedelic hallucinogen obtained from the Peyote cactus). "Under mescaline," he writes, "the sound of a flute gives a bluish-green color, the tick of a metronome, in darkness, is translated as gray patches," while "the barking of a dog is found to attract light in an indescribable way" (Merleau-Ponty 2004 [1962]: 265).

Consensus is yet to be reached within the neurological and psychological research as to the bodily origins and causes of synesthesia. Early speculation was that it is caused by "intertwined nerve fibres" (Calkins 1895: 90); a more recent suggestion is that "cross-wiring" occurs within regions of the brain, or "cross-activation" takes place between brain regions that perform different specialized functions (Ramachandran and Hubbard 2001). Despite the lack of consensus, synesthesia is now understood to be a genuine sensory modality across a spectrum of combinatory forms.

While "cross-modal synesthesia" tends to still be regarded as an anomalous or unusual occurrence or event, there are suggestions that this is not necessarily the case. For neuropsychologist Cytowic, it could well be that synesthetes are more attuned to the holistic nature of sensory experience than most people, but we all have a "capacity for drawing on all our senses to make sense of the world" (Ekdahl 2015: 28). Cytowic (1995b: 206–10) has pointed out that one way we do this – to return to a key thread running through this book – is through our use of metaphor. As Ekdahl explains: "A cheese can be *sharp*, a person might be *sweet*, or a mood can be *low*. Metaphors are wonderful, softcore examples of synesthesia and likely have their roots in the holistic nature of both our perception and cognition" (2015: 28).

Merleau-Ponty goes a step further in suggesting that "synaesthetic perception" is "the rule" not the exception (2004 [1962]: 266). Our body, he argues, is a "synergic system, all the functions of which are exercised and linked together in the general action of being in the world" (272). The issue for Merleau-Ponty is that we have "unlearned" the ability "to see, hear, and ... feel" (266) in sensorially complex ways – in part due to the primacy of technovision, at least in the Western world – and that there remains a need to construct "clearer account[s] of the nature of sensible significance" (267) that capture this complexity.

In its more general usage, the word *synesthesia* captures the "merging of the senses" (Simner 2012). This meaning is conveyed in the word itself, which combines the Greek *syn* (together) with *aesthesis* (perception) (Cytowic 1995a). Steven Connor notes that we do not experience the senses discretely; rather, sensory perception involves "synaesthetic spillings and minglings" (Connor 2004a: 153). He writes:

> The sense we make of any one sense is always mixed with and mediated by that of others. The senses form an indefinite series of integrations and transformations: they form a complexion ... The senses communicate with each other in cooperations and conjugations that are complex, irregular, and multilateral. (2004a: 156)

Of crucial importance in understanding the effect and cultural context of mediating technologies upon sense perception, and the continuing conceptual importance of material metaphor and medium specificity, is how "this complexion of the senses knits itself anew with each new configuration": "What a culture offers is not just a static consortium of the senses, disposed like a molecular structure in a particular configuration, but rather a field of possibility, a repertoire of forms ... Intersensoriality is the means by which this is enacted" (Connor 2004a: 156). In his early work "On Hearing Shapes, Surfaces and Interiors," Ihde (1982) documents the intersensoriality involved in "hearing" the textures, shapes and angularity of objects, noting that sound proximity (such as that detected through reverberation and echo) has a kinesthetic significance (as discussed in the context of mobile haptics and accessibility innovation in Chapter 4). He recounts how his young son was able to quite accurately "hear" the six-sidedness and length of a ball-point pen rolling in a box, and most of us (as we noted in Chapter 3 on ears and listening) are familiar with the action of "knocking" to ascertain density and hollowness. Both Connor (2004a) and Sterne (2006) have also commented on the way Thomas Edison used his teeth to "hear" by biting on the wood of the gramophone to detect faint overtones: "Teeth seem to be involved in the transition from the touched sound of a prerecording era to the untouched sound of a postrecording era. This is because teeth represent an alternative route into the ear or even a way of short-circuiting the ear" (Connor 2004a: 168–9).

In a related way, we can "hear" through our skin the vibration of notifications when our phones, switched to "silent," are placed in pockets close to our body. Indeed, so habituated have we become to this sensory experience that it is not uncommon for mobile phones users to detect and report the sensation of these notificatory vibrations, or "phantom phone sensations," even when their device is not with them (Sauer et al. 2015). In each of the above examples, the perceptual body "conjugates" the senses, and is rendered a membrane or filtering medium as opposed to an ocular apex or point of perspectival origin.

Sensory Ensembles and Mobile Phone Use

How, then, does an understanding of sensory ensembles (intersensoriality) lead us to think differently about our experience of screens, and small mobile screens in particular, especially in ways that skew the visual primacy usually afforded to the screen medium? As we have suggested in this book, the mobile media interface compromises this frontal orientation and visual primacy, as it enacts both separately and in combination visual, haptic, acoustic and mobile incursions into our corporeal schema, and demands variable and oscillating modes of somatic involvement. The mobile phone is an object of corporeal intimacy that far exceeds our visual attachment to the screen. As Leopoldina Fortunati (2005) suggests, the mobile is both multiform and multifunctional, an "open work" requiring a complex range of hermeneutic visual, aural and haptic skills on the part of the user, and is also highly mutable because "it is held very close to the body or stays on the body surface." Even when in "silent" or "vibrate" mode, as noted above, the mobile needs to be in visual, tactile and/or aural proximity, demanding small but constant cross-sensory accommodations and choreographies of the body. At the same time, a notable aspect of the intersensorial embodiment of mobile phones is the manner in which what is seen on the screen merges seamlessly with the hand's movement. In her work on the biomechanical relationship between the hand and mobile screen device, Cooley (2004) describes the tactile vision demanded by mobile screens as a "material and dynamic seeing" which is a collaborative effect of the eyes, hands and device. Thus, the mobile media screen radically skews the stationary and often dedicated relationship which is typical of our engagement with other media screens. Moreover, the eyes and ears of the mobile media device user are constantly distracted by the surrounding clamor and moving objects on the street or sidewalk, by the latent, lateral but ever-ready possibilities of incoming messages, and by the mobility of one's own body in the complex negotiation of urban space. Mobile connectivity is thus rarely a "dedicated" practice. As we have shown throughout this book, it is always-already surrounded by other objects, activities and sounds within the spatial topography and dynamics of the built environment that require overlapping forms of sensory engagement.

The "Era of Cognitive Systems"? Mobile Phones and the Five Senses

In the course of our examination of bodies and mobile media, with its chapter-by-chapter treatment of specific body parts and senses, we have yet to give consideration to gustation (taste) and olfaction (smell). Despite proclamations in the early 2010s that we were entering an "era of cognitive systems" that would lead to a "merger of biological and electronic systems" and the replication of the five human senses on mobile phones (Glenday 2012), the reality is quite different. The gustatory sense, in particular, has proven very difficult to incorporate into mobile devices. The closest developers have managed to come, it would seem, are experiments in "mobile taste," where users in a laboratory setting place electrodes on the tongue that stimulate "virtual taste perception" in the brain, reproducing the sensation of sour, salty, sweet and bitter tastes (Putic 2015).

Olfaction constitutes a somewhat different case, yet needs to be separated into attempts at developing sensors and portable consumer products that are *smell producing* and those that are *smell detecting*. The former (smell production), like taste, has proven difficult to create and, when it has been achieved, difficult to find a market for. In 1999, DigiScents created a USB-connected "scent synthesizer"; despite attracting US$20 million in venture capital funding, by 2001 the company had folded (Twilley 2016). With the arrival of mobile smartphones came Scentee, a mobile app that requires a special dongle-like device with insertable scent cartridges that plugs into the phone's headphone jack (Wee 2014). Based on a user's actions – liking a Facebook post, sending text messages or receiving an alert – the app could be "programmed to release a burst of rose, lavender, or buttered potato scent" (Twilley 2016). In 2014, Vapor Communications created the oPhone, a "large, brick-shaped device mounted with smell-delivery tubes" that was intended to resemble a flower planter. As Twilley (2016) explains, "from a built-in palette of thirty-two scent cartridges," the oPhone "played back oNotes – photographs tagged with up to four smell words, from 'buttery' to 'fishy' to 'yeasty brioche.'" Two years later, Vapor Communications further refined the idea by developing the Cyrano – a "squat, brushed-aluminum cylinder with a perforated lid," marketed as a "digital scent speaker." The Cyrano, Bluetooth-enabled and controlled via a smartphone app, can emit up to twelve different scents. Users of the Cyrano could create "olfactory playlists," much as mobile media users do with Spotify music playlists. Some of these come preloaded: "clicking on 'Thai Beach Vacation,' for example, will play the scents of coconut, suntan lotion, and sea breeze in an infinite loop" (Twilley 2016). Like cinema's infamous Smell-O-Vision, none of these modern-day portable olfactory applications have (as yet) been able to achieve commercial success. From the perspective of the habitual framework of current experience, it seems unlikely that they ever will, not least because smell (like taste) constitutes an "invasion" of the body in a way that other senses do not, an experience many might be averse to. Nevertheless, similar opinions about uptake were expressed about texting (no one will ever use such a small keypad), mobile games (no one will want to play games on a tiny screen) and video-calling (people just want to "talk" on the phone), among other innovations, and each of these have become so integral to our everyday lives as to be mundane.

Meanwhile, developments in "machine olfaction" (smell-detecting mobile sensors and devices) are well underway and have been brought to market with some success. Siemens, for example, has developed mini sensors that can detect particular gasses and other smells (PhysOrg 2004). These sensors have a variety of commercial applications, from quality assurance testing in food and beverage manufacturing, to medical laboratory uses such as testing for airborne harmful bacteria and pathogens (Gardner et al. 2004). UK-based Bullitt Group, which manufactures CAT-branded smartphones, principally targeted for use within construction and other industries, has incorporated scent-detecting sensors into its phones that can monitor for high levels of volatile organic compounds (like glues or paint thinners) and alert the user to their presence (BBC News 2018). Meanwhile, Honeywell has developed the "tricorder," a miniature vacuum pump that serves as an "electronic nose" for use in the US Defense Advanced Research Project Agency's micro-drones. The tricorder is "a full-fledged mobile mass spectrometer that monitors all elements in a particular sample" that can then be run against a database of common substances (Phone Arena 2013). This miniature mass spectrometer could potentially be incorporated into future-generation smartphones.

In addition to machine olfaction, experimental work is also underway exploring the potential of "chemical haptics," which involves the direct application of chemicals to the skin to stimulate various receptors (thermoreceptors, mechanoreceptors, pain receptors, etc.) (Haptics Club 2021). Researchers at the Human Computer Integration Lab at the University of Chicago have developed a wearable sleeve device that employs micropumps to "push liquid stimulants through channels that are open to the user's skin, enabling [these] stimulants to be absorbed by the skin" (Lu et al. 2021). For the wearer of the device, different "haptic sensations" are induced through the application of a variety of "topical stimulants": menthol (a cold sensation), capsaicin (warmth), lidocaine (numbing), sanshool (tingling), cinnamaldehyde (stinging) and methyl salicylate (hot and cold at the same time) (Lu et al. 2021). One application of chemical haptics is in gaming, with the Lab's researchers connecting the sleeve to a VR headset. When the user "shoots electrical rays from a 'portable tesla coil' on their arm," the act of shooting triggers a tingling sensation, which the sleeve device renders by administering sanshool onto the skin of the forearm (Lu et al. 2021). While such innovations, in extending the capacity of "touch" technology by permeating the boundary of the skin, raise a host of ethical issues, they also point to the ways in which touch is a "manifold" sense in Paterson's (2007) terms, and thus inherently open to synesthetic effects.

Multispecies Body–Technology Couplings

These attempts at expanded sensory engagement provide a useful point-of-entry for thinking about emerging work on corporeality and body–technology couplings, especially as they relate to critical work on more-than-human perception and embodied, multispecies encounters with screen media.

Across the social sciences, for instance, there has been growing awareness and strengthening of critical interest in sensory-focused modes of enquiry (e.g. Pink 2015b); increased attentiveness to more-than-human and multispecies ways of understanding and engaging with the world (e.g. Galloway 2017; Puig de la Bellacasa 2017; Haraway 2003, 2008; Eben Kirksey and Helmreich 2010; Whatmore 2006) – not at all surprising given that "there is no sense which humanity does not share with the animal world" (Barthes 1991: 246); and growing attention paid to how technological knowledges can be shared across species boundaries (Shew 2017).

In *On Looking*, Alexandra Horowitz takes experts – including a geologist, a typographer and a naturalist – on a series of twelve walks around New York City to understand what they see that "amateur eyes" (2013: 1) might not. On her final walk, Horowitz takes her own dog. What she recalled about her previous dog Pumpernickel, and noticed in her present dog Finn, was that "she inhaled her view with her nose" (242). Dogs are "macrosomatic" (keen-scented) and can "detect odors at one or two parts per trillion" (245). Horowitz observed that, just as we might visually scan a city scene, Finn smelled the scene. This is because scents are not static or unchanging: "They are a haze, a cloud, spreading out from their source" (248). An olfactory encounter with a street is experienced as "a mishmash of overlapping object identities, each crowding into the next's odorous space" (249). The ability of dogs to independently sample odors in each nostril leads Horowitz to suggest that this "may allow dogs a kind of stereo olfaction" (249). Whether or not this is so, it remains the case that for dogs "their world has a topography wrought of odors; the landscape is brightly colored with aromas" (244).

Recent work within media studies has explored how companion animals (Richardson et al. 2017), primates (Webber et al. 2017) and other biological organisms come to form "entanglements of media, body, and environment" (Mukherjee 2020), especially in the context of pet wearables and webcams used to monitor the home. Not only do these lines of enquiry lead us to understand that all "those who are in the world are constituted in intra- and interaction" – that we are all "beings-in-encounter" (Haraway 2008: 4, 5) – they also help us to comprehend "a variety of other embodied intentionalities" that can reveal "new ways of kinesthetically and corporeally interacting" with mobile media devices (Richardson et al. 2017: 114). Such work prompts us to further think through possible futures that are more ethically attuned to multispecies agency beyond the Anthropocene.

Summary

This book has sought to translate the "corporeal turn" and the postphenomenological framework in the context of mediated perception, and to explore the close relation between bodies, sensory perception and media use over the past few decades. All media combine sensory magnification with sensory diminishment – every medium is a means of sensory filtering or a conduit to sensory affordance. We have focused on the corporeal schematics particular to our use of mobile media, primarily exploring the relation between human bodies and mobile phones in terms of techno-somatic involvement, and recognizing the medium specificity of our corporeal schema in relation to technologies-in-use. Across the book's central chapters, we first examined how the face–screen metaphorically and physically orientates our use of mobile media, and then explored its visual, aural, tactile and pedestrian aspects in turn.

We have considered how the facial ontology of screens, including mobile screens though to a lesser extent than large stationary screens, reinforces a certain material "kinship" with faces, determining degrees of attention and an intimacy reflected in the rise of the selfie as an important "affective" moment in media ecology. In an historical context, we documented that while many technologies have a perceptual bias towards vision, mobile media practices and contemporary "plurivision" (Ihde 2002) reveal a certain "loosening" of the eye's attachment to the screen, as haptics and the distraction of pedestrian mobility take key sensory roles. Attuning ourselves to the phenomenology of listening and hearing, we examined how mobile media are co-opted in enactments of privacy, transforming our behavior in urban space, and deployed as listening devices for capturing personal information and micro-data. Of perhaps the most significance in terms of recent innovations, the mobile touchscreen has heralded a revolution in mediated haptic perception that has worked to ameliorate the isolating impacts of the pandemic, yet it often remains the case that "ableist" design and development in mobile haptics marginalizes the multistable experience of synesthetic perception across culture and embodied capacity. Mobile media are also fundamentally *mobile*, thus a focus on pedestrian or peripatetic mobile practices is of key importance, especially in terms of how the mobile interface modifies our habitual ambulatory activities, and how navigational and playful apps both augment and redirect our trajectories through the urban environment.

As noted in our discussion of cross-modal experience, the separation of these sense modalities does not imply that they are in fact separable, and neither has it been our intention to suggest that the formal five-sense structure should be taken as fact. As Ward (2023) argues, such reductionism does a great disservice to the rich variety of our sensory experience, with many scholars suggesting that we have twenty senses or more, including proprioception (bodily awareness of posture and position in relation to space), kinesthesia (bodily awareness of the extent and direction of movement of limbs), equilibrioception (balance), mechanoreception (acceleration and muscle stretch), interoception (perception of our inner biology, such as a full stomach and lung inflation), not to mention the myriad of "affective senses" (sense of self, emotion, mood and so on). We have addressed some of these extra-sensory modalities throughout the book, though of course there are many we have not discussed, as their extraordinary variation and constant mobile innovation make this project an ever-evolving and open one.

Forty years ago, Robert Innis quite radically suggested that our current technologies have "provided us with the most controversial 'body' in the history of humanity" (1984: 69), a claim that resonates ever more deeply in the contemporary context. With this incisive proposition in mind, our work has proposed new ways of describing the expandable and pliable nature of our corporeal schema, alternative metaphors for defining existing and emerging modes of embodiment afforded by mobile media interfaces, and more adaptive conceptual approaches for understanding how our bodily boundaries and experiences are continually reinvented by technological mediation.

References

Adams, M. L. (2019). Step-counting in the "health-society": Phenomenological reflections on walking in the era of the Fitbit. *Social Theory and Health*, 17(1), pp. 109–24.

Alderman, N. (2019). Turn your run into a thriller. Behind the Scenes. Apple App Store, https://apps.apple.com/nz/story/id1297896684

Amato, J. (2004). *On Foot: A History of Walking*. New York: New York University Press.

Andrejevic, M. (2015). Becoming drones: Smartphone probes and distributed sensing. In R. Wilken and G. Goggin, eds. *Locative Media*. New York: Routledge, pp. 193–207.

ANU TV (2017). How does the crested pigeon make their mysterious alarm sound? November 10, https://youtu.be/6tKyG6-zaBo

Apperley, T. and Moore, K. (2019). Haptic ambiance: Ambient play, the haptic effect and co-presence in *Pokémon GO*. *Convergence: The International Journal of Research into New Media Technologies*, 25(1), pp. 6–17.

Ardiel, E. L. and Rankin, C. H. (2010). The importance

of touch in development. *Paediatrics & Child Health*, 15(3), pp. 153–6, https://doi.org/10.1093/pch/15.3.153

Arnold, M. (2003). On the phenomenology of technology: The "Janus-faces" of mobile phones. *Information and Organization*, 13(4), pp. 231–56.

Barad, K. (2012). On touching – the inhuman that therefore I am. *differences: A Journal of Feminist Cultural Studies*, 23(3), pp. 206–23.

Barnett, K. J., Finucane, C., Asher, J. E., Bargary, G., Corvin, A. P., Newell, F. N. and Mitchell, K. J. (2008). Familial patterns and the origins of individual differences in synaesthesias. *Cognition*, 106, pp. 871–93, https://doi.org/10.1016/j.cognition.2007.05.003

Baron, N. S. (2010). *Always On: Language in an Online and Mobile World*. Oxford: Oxford University Press.

Barthes, R. (1991). Listening. In *The Responsibility of Forms: Critical Essays on Music, Art, and Representation*. Trans. R. Howard. Berkeley: University of California Press, pp. 245–60.

Bassett, C. (2005). How many movements? Mobile telephones and transformations in urban space. *Open! Platform for Art, Culture & the Public Domain*, https://onlineopen.org/how-many-movements

BBC News (2018). Smell-sensing smartphone sniffs out glue. *YouTube*, February 25, https://youtu.be/28kSWKu_Uzk

Bitman, N. and John, N. A. (2019). Deaf and hard of hearing smartphone users: Intersectionality and the penetration of ableist communication norms. *Journal of Computer-Mediated Communication*, 24, pp. 56–72.

Brown, N. O. (1966). *Love's Body*. New York: Random House.

Brune, A. and Wilson, D. J. (2013). Introduction. In A. Brune and D. J. Wilson, eds. *Disability and Passing:*

Blurring the Lines of Identity. Philadelphia, PA: Temple University Press, pp. 1–12.

Bull, M. (2004a). Thinking about sound, proximity and distance in Western experience: The case of Odysseus's Walkman. In V. Erlmann, ed. *Hearing Cultures: Essays on Sound, Listening and Modernity*. Oxford: Berg, pp. 173–90.

Bull, M. (2004b). Sound connections: An aural epistemology of proximity and distance in urban culture. *Environment and Planning D: Society and Space*, 22, pp. 103–16.

Bull, M. (2004c). "To each their own bubble": Mobile spaces of sound in the city. In N. Couldry and A. McCarthy, eds. *MediaSpace: Place, Scale and Culture in a Media Age*. London: Routledge, pp. 275–93.

Bull, M. (2007). *Sound Moves: iPod Culture and Urban Experience*. London: Routledge.

Burgess, J., Albury, K., McCosker, A. and Wilken, R. (2022). *Everyday Data Cultures*. Cambridge: Polity Press.

Cacciari, C. (2008). Crossing the senses in metaphorical language. In R. W. Gibbs, ed. *The Cambridge Handbook of Metaphor and Thought*. Cambridge: Cambridge University Press, pp. 425–43.

Calkins, M. W. (1895). Synaesthesia. *The American Journal of Psychology*, 7(1), pp. 90–107.

Campanella, T. J. (2000). Eden by wire: Webcameras and the telepresent landscape. In K. Goldberg, ed. *The Robot in the Garden: Telerobotics and Telepistemology in the Age of the Internet*. Cambridge, MA: MIT Press, pp. 22–46.

Campbell, S. (2008). Perceptions of mobile phone use in public: The roles of individualism, collectivism, and focus of the setting. *Communication Reports*, 21(2), pp. 70–81.

Canetti, Elias (1962). *Crowds and Power*. New York: Farrar, Straus and Giroux.

Capron, A. (2018). Did this cat really cuddle up to a video of his former owner – or is it a hoax? *The Observers*, September 17, https://observers.france24.com/en/20180917-cat-cuddle-video-dead-owner-hoax

Carey, J. (1969). Harold Adams Innis and Marshall McLuhan. In R. Rosenthal, ed. *McLuhan: Pro and Con*. Baltimore, MD: Pelican Baltimore, pp. 270–308.

Cashell, G. T. W. (1971). A short history of spectacles. *Proceedings of the Royal Society of Medicine*, 64, pp. 1063–4.

Chan, C. W. and Rudins, A. (1994). Foot biomechanics during walking and running. *Mayo Clinic Proceedings*, 69(5), pp. 448–61.

Chesher, C. (2004). Neither gaze nor glance, but glaze: Relating to console game screens. *SCAN: Journal of Media Arts Culture*, 1(1), http://scan.net.au/scan/journal/display.php?journal_id=19

Choi, J. (2007). Approaching the mobile culture of East Asia. *M/C Journal*, 10(1), http://journal.media-culture.org.au/0703/01-choi.php

Cocozza, P. (2018). No hugging: Are we living through a crisis of touch? *The Guardian*, March 18, https://www.theguardian.com/society/2018/mar/07/crisis-touch-hugging-mental-health-strokes-cuddles

Colley, A., Thebault-Spieker, J., Yilun Lin, A., Degraen, D., Fischman, B., Hakkila, J., Kuehl, K., Nisi, V., Nunes, N. J., Wenig, N., Wenig, D., Hecht, B. and Schöning, J. (2017). The geography of Pokémon GO: Beneficial and problematic effects on places and movement. *CHI 2017*, 6–11 May, Denver, CO.

Connor, S. (1999). Remarks on music and listening. In M. Oliver, ed. *Settling the Score: Journey Through Music of the Twentieth Century*. London: Faber and Faber, pp. 144–205.

Connor, S. (2004a). Edison's teeth: Touching hearing. In V. Erlmann, ed. *Hearing Cultures: Essays on Sound, Listening and Modernity*. Oxford and New York: Berg, pp. 153–72.

Connor, S. (2004b). Sound and the self. In M. M. Smith, ed. *Hearing History: A Reader*. Athens: University of Georgia Press, pp. 54–66.

Coole, D and Frost, S. (eds) (2010). *New Materialisms: Ontology, Agency, and Politics*. Durham, NC: Duke University Press.

Cooley, H. R. (2004). It's all about the *fit*: The hand, the mobile screenic device, and tactile vision. *Journal of Visual Culture*, 3(2), pp. 133–55.

Cooley, H. R. (2014). *Finding Augusta: Habits of Mobility and Governance in the Digital Era*. Lebanon, NH: University Press of New England.

Couldry, N. (2012). *Media, Society, World: Social Theory and Digital Media Practice*. Cambridge: Polity Press.

Crawford, A. and Goggin, G. (2009). Geomobile web: Locative technologies and mobile media. *Australian Journal of Communication*, 26(1), pp. 97–109.

Crawford, K. (2012). Four ways of listening with an iPhone: From sound and network listening to biometric data and geolocative tracking. In L. Hjorth, J. Burgess and I. Richardson, eds. *Studying Mobile Media: Cultural Technologies, Mobile Communication and the iPhone*. New York: Routledge, pp. 213–28.

Crawford, K., Lingel, J. and Karppi, T. (2015). Our metrics, ourselves: A hundred years of self-tracking from the weight scale to the wrist wearable device. *European Journal of Cultural Studies*, 18(4–5), pp. 479–96.

Cytowic, R. E. (1995a). Synesthesia: Phenomenology and neuropsychology. A review of current knowledge.

Psyche, 2(10), https://journalpsyche.org/files/0xaa34.pdf

Cytowic, R. E. (1995b). *The Man Who Tasted Shapes*. New York: Warner Books.

Daxecker, F. (1997). Representations of eyeglasses on Gothic winged altars in Austria. *Documenta Ophthalmologica*, 93, pp. 169–88.

de Certeau, M. (1984). *The Practice of Everyday Life*. Trans. S. Rendall. Berkeley: University of California Press.

de Souza e Silva, A. (2004). Mobile networks and public spaces: Bringing multiuser environments into the physical space. *Convergence: The International Journal of Research into New Media Technologies*, 10(2), pp. 15–25.

de Zengotita, T. (2014). We love screens, not Glass. *The Atlantic*, March, http://www.theatlantic.com/technology/archive/2014/03/we-love-screens-not-glass/284356

DeSilva, J. (2021). *First Steps: How Walking Upright Made Us Human*. London: HarperCollins.

Do The Right Thing (1989). Directed by Spike Lee [film]. Universal Pictures.

Donald, S. H. and Richardson, I. (2002). The English project: Function and culture in new media research. *Inter/Sections: The Journal of Global Communications and Culture*, 2(5), pp. 155–66.

Duggan, M. (2017). The cultural life of maps: Everyday place-making mapping practices. *Living Maps Review*, 3, http://livingmaps.review/journal/index.php/LMR/article/view/71/119

Duggan, M. (2018). Navigational mapping practices: Contexts, politics, data. *Westminster Papers in Communication and Culture*, 13(2), pp. 31–45, https://doi.org/10.16997/wpcc.288

Dunham, J., Papangelis, K., Laato, S., Lalone, N., Lee, J. H. and Saker, M. (2022). *Pokémon GO* to *Pokémon STAY*: How COVID-19 affected *Pokémon GO* players. *ACM Transactions in Computer-Human Interaction*, 1(1), https://arxiv.org/abs/2202.05185

Eben Kirksey, S. and Helmreich, S. (2010). The emergence of multispecies ethnography. *Cultural Anthropology*, 25(4), pp. 545–76.

Eckert, R. C. and Rowley, A. J. (2013). Audism: A theory and practice of audiocentric privilege. *Humanity & Society*, 37(2), pp. 99–193.

Eikenes, J.O.H. and Morrison, A. (2010). Navimation: Exploring time, space and motion in the design of screen-based interfaces. *International Journal of Design*, 4(1), pp. 1–16.

Ekdahl, D. (2015). Holism in perception: Merleau-Ponty on senses and objects. *TIDskrift*, VII, pp. 22–37.

Elo, M. (2012). Formatting the senses of touch. *Transformations*, 22, pp. 1–15, http://www.transformationsjournal.org/wp-content/uploads/2016/12/Elo_Trans22.pdf

Evans, L. (2015). *Locative Social Media: Place in the Digital Age*. Basingstoke, Hampshire: Palgrave Macmillan.

Evans, L. and Saker, M. (2017). *Location-Based Social Media: Space, Time and Identity*. Cham, Switzerland: Palgrave Macmillan.

Fabiny, T. (2005). The ear as a metaphor: Aural imagery in Shakespeare's great tragedies and its relation to music and time in "Cymbeline" and "Pericles." *Hungarian Journal of English and American Studies (HJEAS)*, 11(1), pp. 189–201.

Farman, J. (2012). *Mobile Interface Theory: Embodied Space and Locative Media*. London: Routledge.

Farman. J. (2015). Stories, spaces, and bodies: The production of embodied space through mobile media storytelling. *Communication Research and Practice*, 1(2), pp. 101–16.

Farris, D. J., Kelly, L. A., Cresswell, A. G. and Lichtwark, G. A. (2019). The functional importance of human foot muscles for bipedal locomotion. *PNAS*, 116(5), pp. 1645–50.

Ferguson, K. (2008). Aural addictions. *Continuum: Journal of Media and Cultural Studies*, 22(1), pp. 69–77.

Fischer, C. S. (1992). *America Calling: A Social History of the Telephone to 1940*. Berkeley: University of California Press.

Fleming, A. (2019). "It's a superpower": How walking makes us healthier, happier, brainier. *The Guardian*, July 29, https://www.theguardian.com/lifeandstyle/2019/jul/28/its-a-superpower-how-walking-makes-us-healthier-happier-and-brainier

Fortunati, L. (2005). The mobile phone as technological artefact. In P. Glotz, S. Bertschi and C. Locke, eds. *Thumb Culture: The Meaning of Mobile Phones for Society*. Bielefeld: transcript Verlag, pp. 149–60.

Friedberg, A. (2006). *The Virtual Window: From Alberti to Microsoft*. Cambridge, MA: MIT Press.

Frith, J. (2014). Communicating through location: The understood meaning of the Foursquare check-in. *Journal of Computer-Mediated Communication*, 19, pp. 890–905.

Frith, J. (2020). RFID, NFC, beacons and the infrastructures of logistical locative media. In R. Ling, L. Fortunati, G. Goggin, S. S. Lim and Y. Li, eds. *The Oxford Handbook of Mobile Communication and Society*. New York: Oxford University Press, pp. 475–86.

Frith, J. and Kalin, J. (2016). "Here, I used to be": Mobile media and practices of place-based digital memory. *Space and Culture*, 19(1), pp. 43–55.

Frosh, P. (2015). The gestural image: The selfie, photography theory, and kinesthetic sociability. *International Journal of Communication*, 9, pp. 1607–28.

Galloway, A. (2013). Emergent media technologies, speculation, expectation, and human/nonhuman relations. *Journal of Broadcast & Electronic Media*, 57(1), pp. 53–65.

Galloway, A. (2017). More-than-human lab: Creative ethnography after human exceptionalism. In L. Hjorth, H. Horst, A. Galloway and G. Bell, eds. *The Routledge Companion to Digital Ethnography*. New York: Routledge, pp. 470–7.

Gandelman, C. (1991). *Reading Pictures, Viewing Texts*. Bloomington, IN: Indiana University Press.

Gardner, J., Hines, E. and Dowson, C. (2004). Smelling illness. *Ingenia*, 21, pp. 38–9, http://www.ingenia.org.uk/Ingenia/Issue-21/Smelling-illness

Gergen, K. (2002). The challenge of absent presence. In J. E. Katz and M. Aakhus, eds. *Perpetual Contact: Mobile Communication, Private Talk, Public Performance*. Cambridge: Cambridge University Press, pp. 227–41.

Gibson, J. J. (1986). *The Ecological Approach to Visual Perception*. Hillsdale, NJ: Lawrence Earlbaum Associates.

Gilmore, J. N. (2016). Everywear: The quantified self and wearable fitness technologies. *New Media & Society*, 18(11), pp. 2524–39.

Glenday, J. (2012). IBM to replicate the five senses of humans on mobile phones. *The Drum*, December 18, https://www.thedrum.com/news/2012/12/18/ibm-replicate-five-senses-mobile-phone

Goffman, E. (1963). *Behavior in Public Places: Notes on the Social Organization of Gatherings*. New York: The Free Press.

Goffman, E. (1967). *Interaction Ritual: Essays on Face-to-Face Behavior*. New York: Pantheon Books.

Goffman, E. (1972). *Relations in Public: Microstudies of the Public Order*. Harmondsworth: Penguin.

Goggin, G. (2016). Disability and mobilities: Evening up social futures. *Mobilities*, 11(4), pp. 533–41.

Goggin, G. (2017). Disability and haptic mobile media. *New Media & Society*, 19(10), pp. 1563–80.

Goggin, G. and Hjorth, L. (2009). The question of mobile media. In G. Goggin and L. Hjorth, eds. *Mobile Technologies: From Telecommunications to Media*. New York: Routledge, pp. 3–8.

Goode, L. (2021). Augmented reality is coming for your ears, too. *Wired*, August 13, https://www.wired.com/story/augmented-reality-already-arrived-in-our-ears

Gopinath, S. (2013). *The Ringtone Dialectic: Economy and Cultural Form*. Cambridge, MA: MIT Press.

Griffiths, L. E. (2023). Dancing through social distance: Connectivity and creativity in the online space. *Body, Space & Technology*, 22(1), pp. 65–81.

Gu, C. and Griffin, M. J. (2011). Vibrotactile thresholds at the sole of the foot: Effect of vibration frequency and contact location. *Somatosensory & Motor Research*, 28(3–4), pp. 86–93.

Hall, J. (2004). Mogi: Second generation location-based gaming. *The Feature*, April 1, https://web.archive.org/web/20180215105156/http://links.net/share/write/thefeature/Mogi__Second_Generation_Location-Based_Gaming.html

Hammer, A. (2007). Audible evidence: On listening to places. *Jump Cut: A Review of Contemporary Media*, 49 (Spring), https://www.ejumpcut.org/archive/jc49.2007/HammerAudio

Haptics Club. (2021). Feeling touch and temperature with chemical haptics. *Haptics Club* [podcast], November 23, https://podcasters.spotify.com/pod/show/haptics

-club/episodes/13-Feeling-Touch-and-Temperature-with-Chemical-Haptics-e1ammne/a-a6v4fr7
Haraway, D. J. (1988). Situated knowledges: The science question in feminism and the privilege of partial perspective. *Feminist Studies*, 14(3), pp. 575–99.
Haraway, D. J. (1991). *Simians, Cyborgs and Women: The Reinvention of Nature*. New York: Routledge.
Haraway, D. (2003). *The Companion Species Manifesto: Dogs, People, and Significant Otherness*. Chicago: Prickly Paradigm Press.
Haraway, D. (2008). *When Species Meet*. Minneapolis: University of Minnesota Press.
Harding, S. and Hintikka, M. B., eds. (1983). *Discovering Reality: Feminist Perspectives on Epistemology, Metaphysics, Methodology, and Philosophy of Science*. Dordrecht, Holland: D. Reidel.
Hardley, J. and Richardson, I. (2021). Mistrust of the city at night: Networked connectivity and embodied perceptions of risk and safety. *Australian Feminist Studies*, 36(107), pp. 65–81, https://doi.org/10.1080/08164649.2021.1934815
Harmon, K. (2013). Growing up to become hearing: Dreams of "passing" in oral deaf education. In A. Brune and D. J. Wilson, eds. *Disability and Passing: Blurring the Lines of Identity*. Philadelphia, PA: Temple University Press, pp. 167–98.
Heelan, P. (1983). *Space Perception and the Philosophy of Science*. Berkeley: University of California Press.
Heidegger, M. (1977). *The Question Concerning Technology and Other Essays*. New York: Garland Publishing, Inc.
Helmenstine, A. M. (2019). Human eye anatomy. December 2, https://www.thoughtco.com/how-the-human-eye-works-4155646
Helyer, N. (2007). The sonic commons: Embrace or

retreat? *Scan Journal*, 4(3), http://scan.net.au/scan/journal/display.php?journal_id=105

Henning, E. M. and Sterzing, T. (2009). Sensitivity mapping of the human foot: Thresholds at 30 skin locations. *Foot & Ankle International*, 30(10), pp. 986–91.

Higuchi, T. (1983). *The Visual and Spatial Structure of Landscape*. Trans. C. S. Terry. Cambridge, MA: MIT Press.

Hjorth, L. (2009). *Mobile Media in the Asia-Pacific: Gender and the Art of Being Mobile*. New York: Routledge.

Hjorth, L. and Pink, S. (2014). New visualities and the digital wayfarer: Reconceptualizing camera phone photography and locative media. *Mobile Media & Communication*, 2(1), pp. 40–57.

Hjorth, L. and Richardson, I. (2009). The waiting game: Complicating notions of (tele)presence and gendered distraction in casual mobile gaming. *Australian Journal of Communication*, 36(1), pp. 23–35.

Hjorth, L. and Richardson, I. (2020). *Ambient Play*. Cambridge, MA: MIT Press.

Hjorth, L., Richardson, I., Davies, H. and Balmford, W. (2020). *Exploring Minecraft: Ethnographies of Play and Creativity*. Cham, Switzerland: Palgrave Macmillan.

Höflich, J. R. (2005). A certain sense of place: Mobile communication and local orientation. In K. Nyíri, ed. *A Sense of Place: The Global and the Local in Mobile Communications*. Vienna: Passagen Verlag, pp. 159–68.

Horowitz, A. (2013). *On Looking: About Everything There Is to See*. London: Simon and Schuster.

Hosokawa, S. (1984). The Walkman effect. *Popular Music*, 4, pp. 165–80.

Ihde, D. (1979). *Technics and Praxis*. Dordrecht: D. Riedel Publishing Company.

Ihde, D. (1982). On hearing shapes, surfaces and interiors.

In R. Bruzina and B. Wilshire, eds. *Phenomenology: Dialogues and Bridges*. Albany, NY: State University of New York Press, pp. 241–52.

Ihde, D. (1990). *Technology and the Lifeworld: From Garden to Earth*. Bloomington, IN: Indiana University Press.

Ihde, D. (1991). *Instrumental Realism: The Interface Between Philosophy of Science and Philosophy of Technology*. Bloomington, IN: Indiana University Press.

Ihde, D. (1993). *Postphenomenology: Essays in the Postmodern Context*. Evanston, IL: Northwestern University Press.

Ihde, D. (2002). *Bodies in Technology*. Minneapolis: University of Minnesota Press.

Ihde, D. (2007). *Listening and Voice: Phenomenologies of Sound*, second edition. Albany, NY: State University of New York Press.

Ihde, D. (2010). A phenomenology of technics. In C. Hanks, ed. *Technology and Values: Essential Readings*. Malden, MA: Wiley-Blackwell, pp. 134–55.

Ingold, T. (2004). Culture on the ground: The world perceived through the feet. *Journal of Material Culture*, 9(3), pp. 315–40.

Ingold, T. (2007). *Lines: A Brief History*. London: Routledge.

Ingold, T. (2010). Footprints through the weather-world: Walking, breathing, knowing. *Journal of the Royal Anthropological Institute*, 16(s1), pp. S121–S139.

Ingold, T. (2011). *The Perception of the Environment: Essays on Livelihood, Dwelling and Skill*. London: Routledge.

Innis, H. (1964). *The Bias of Communication*. Toronto: University of Toronto Press.

Innis, R. E. (1984). Technics and the bias of perception. *Philosophy and Social Criticism*, 10(1), pp. 67–89.

InTouch Digital (2020). Media tales of touch during Covid. April 23, https://in-touch-digital.com/2020/04/23/media-tales-of-touch-during-covid

Introna, L. D. and Ilharco, F. M. (2004). The ontological screening of contemporary life: A phenomenological analysis of screens. *European Journal of Information Systems*, 13, pp. 221–34.

Irwin, S. O. (2016). *Digital Media: Human–Technology Connection*. Lanham, MD: Lexington Books.

Ito, M. (2003). Mobiles and the appropriation of place. *Receiver*, 8. Archived at https://archives.evergreen.edu/webpages/curricular/2003-2004/evs/readings/itoShort.pdf

Ito, M. (2005). Mobile phones, Japanese youth, and the re-placement of social contact. In R. Ling and P. E. Pedersen, eds. *Mobile Communications: Renegotiations of the Social Sphere*. London: Springer, pp. 131–48.

Ito, M. and Okabe, D. (2005). Technosocial situations: Emergent structurings of mobile email use. In M. Ito, D. Okabe and M. Matsuda, eds. *Personal, Portable, and Pedestrian: Mobile Phones in Japanese Life*. Cambridge, MA: MIT Press, pp. 257–73.

Jack, R. E. and Schyns, P. G. (2015). The human face as a dynamic tool for social communication. *Current Biology*, 25(14), https://www.sciencedirect.com/science/article/pii/S0960982215006557

Jacob, R. J. K., Girouard, A., Hirshfield, L. M., Horn, M. S., Shaer, O., Solovey, E. T. and Zigelbaum, J. (2008). Reality-based interaction: A framework for post-WIMP interfaces. *CHI 2008 Proceedings of the SIGCHI conference on human factors in computing systems*, Florence, Italy, April 5–10, New York: ACM, pp. 201–10.

Jaworski, Ł. and Karpiński, R. (2017). Biomechanics of the

human hand. *Journal of Technology and Exploitation in Mechanical Engineering*, 3(1), pp. 28–33.

Jenks, C., ed. (1995). *Visual Culture*. London: Routledge.

Jethani, S. (2021). *The Politics and Possibilities of Self-Tracking Technology*. Bingley: Emerald.

Jewitt, C., Price, S., Steimle, J., Huisman, G., Golmohammadi, L., Pourjafarian, N., Frier, W., Howard, T., Ipakchian Askar, S., Ornati, M., Panëels, S. and Weda, J. (2021). Manifesto for digital social touch in crisis. *Frontiers in Computer Science*, 3, 754050, https://doi.org/10. 3389/fcomp.2021.754050

Jones, E. G. (2006). The sensory hand. *Brain*, 129(12), pp. 3413–20.

Kale, S. (2020). Skin hunger helps explain your desperate longing for human touch. *Wired*, April 29, https://www.wired.co.uk/article/skin-hunger-coronavirus-human-touch

Keller, E. F. and Grontkowski, C. (1983). The mind's eye. In S. Harding and M. Hintikka, eds. *Discovering Reality: Feminist Perspectives and Epistemology, Metaphysics, Methodology and the Philosophy of Science*. Dordrecht: Reidel, pp. 207–44.

Kenaan, H. (2018). The selfie and the face. In J. Eckel, J. Ruchatz and S. Wirth, eds. *Exploring the Selfie: Historical, Theoretical, and Analytical Approaches to Digital Self-Photography*. Cham, Switzerland: Palgrave Macmillan.

Keogh, B. (2015). *A Play of Bodies: A Phenomenology of Videogame Experience*. PhD dissertation. RMIT University, Melbourne, Australia.

Keogh, B. (2018). *A Play of Bodies*. Cambridge, MA: MIT Press.

Kirkpatrick, G. (2009). Controller, hand, screen: Aesthetic form in the computer game. *Games and Culture*, 4(2), pp. 127–43.

Kopomaa, T. (2000). *The City in Your Pocket: Birth of the Mobile Information Society*. Helsinki: Gaudeamus.
Koziol, M. (2015). How Apple Watch changed Molly's life. *Sydney Morning Herald*, May 5, https://www.smh.com.au/technology/how-apple-watch-changed-mollys-life-20150505-ggu4hu.html
Kuipers, J. C. and Bell, J. A. (2018). Introduction: Linguistic and material intimacies of cell phone communication. In J. A. Bell and J. C. Kuipers, eds. *Linguistic and Material Intimacies of Cell Phones*. London: Routledge, pp. 1–30.
Lacoma, T. and Cohen, S. (2022). What is spatial audio? Apple's 3D sound feature fully explained. *Digitaltrends*, August 16, https://www.digitaltrends.com/home-theater/apples-spatial-audio-explained
Lakoff, G. and Johnson, M. (1980). *Metaphors We Live By*. Chicago, IL: University of Chicago Press.
Lakoff, G. and Johnson, M. (1999). *Philosophy in the Flesh: The Embodied Mind and its Challenge to Western Thought*. New York: Basic Books.
Lasen, A. (2004). Affective technologies – emotions and mobile phones. *receiver*, 11, Archived at https://robertoigarza.files.wordpress.com/2009/07/art-affective-technologiese28093emotionsmobile-phones-lasen-2006.pdf
Lasen, A. (2006). How to be in two places at the same time? Mobile phone use in public places. In J. Höflich and M. Hartmann, eds. *Mobile Communication in Everyday Life: Ethnographic Views, Observations and Reflections*. Berlin: Frank & Timme, pp. 227–51.
Lasen, A. (2018). Disruptive ambient music: Mobile phone music listening as portable urbanism. *European Journal of Cultural Studies*, 21(1), pp. 96–110.
Leder, D. (1990). *The Absent Body*. Chicago, IL: University of Chicago Press.

Lee, D. N. and Lishman, J. R. (1975). Visual proprioceptive control of stance. *Journal of Human Movement Studies*, 1(2), pp. 87–95, https://psycnet.apa.org/record/1976-27111-001

Lefebvre, H. (1974). *The Production of Space*. Trans. D. Nicholson-Smith. Malden, MA: Editions Anthropos.

Lefebvre, H. (2004). *Rhythmanalysis: Space, Time and Everyday Life*. Trans. S. Elden and G. Moore. London: Continuum.

Leonard, J. (2022). 5 reasons why Google Glass was a miserable failure, https://www.business2community.com/tech-gadgets/5-reasons-google-glass-miserable-failure-01462398

Licoppe, C. (2004). "Connected" presence: The emergence of a new repertoire for managing social relationships in a changing communication technoscape. *Environment and Planning D: Society and Space*, 22(1), pp. 135–56.

Licoppe, C. (2010a). The "crisis of the summons": A transformation in the pragmatics of "notifications," from ring tones to instant messaging. *The Information Society*, 26(4), pp. 288–302.

Licoppe, C. (2010b). What does answering the phone mean? A sociology of the phone ring and musical ringtones. *Cultural Sociology*, 5(3), pp. 367–84.

Licoppe, C. and Inada, Y. (2006). Emergent uses of a multiplayer location-aware mobile game: The interactional consequences of mediated encounters. *Mobilities*, 1(1), pp. 39–61.

Licoppe, C. and Inada, Y. (2008). Geolocalized technologies, location aware communities and personal territories: The Mogi case. *Journal of Urban Technology*, 15(3), pp. 5–24.

Licoppe, C. and Inada, Y. (2010). Locative media and cultures of mediated proximity: The case of the Mogi game location-aware community. *Environment and Planning D: Society and Space*, 28, pp. 691–709.

Light, A. (2009). Negotiations in space: The impact of receiving phone calls on the move. In R. Ling and S. W. Campbell, eds. *The Reconstruction of Space and Time: Mobile Communication Practices*. New Brunswick, NJ: Transaction Publishers, pp. 191–213.

Ling, R. (2002). The social juxtaposition of mobile telephone conversations and public spaces. In *The Social and Cultural Impact/Meaning of Mobile Communication*, Chunchon Conference on Mobile Communication, July 13–14, 2002, pp. 59–86.

Ling, R. and Haddon, L. (2003). Mobile telephony, mobility and the coordination of everyday life. In J. Katz, ed. *Machines That Become Us: The Social Context of Personal Communication Technology*. New Brunswick, NJ: Transaction Publishers, pp. 245–65.

Lord, S. R., Dayhew, J. and Howland, A. (2002). Multifocal glasses impair edge-contrast sensitivity and depth perception and increase the risk of falls in older people. *Journal of the American Geriatric Society*, 50(11), pp. 1760–6.

Lu, J., Liu, Z., Brooks, J. and Lopes, P. (2021). Chemical haptics: Rendering haptic sensations via topical stimulants. *UIST '21: The 34th Annual ACM Symposium on User Interface Software and Technology*, Virtual Event, USA, October 2021, https://doi.org/10.1145/3472749.3474747

Lupton, D. (2016). *The Quantified Self*. Cambridge: Polity.

McCartney, A. (2014). Soundwalking: Creating moving environmental sound narratives. In S. Gopinath and J. Stanyek, eds. *The Oxford Handbook of Mobile Music Studies, Volume 2*. New York: Oxford University Press, pp. 212–37.

McCosker, A. and Wilken, R. (2020). *Automating Vision:*

The Social Impact of the New Camera Consciousness. New York: Routledge.

McLuhan, M. (2003 [1964]). *Understanding Media*. London: Routledge.

McQuire, S. (2003). From glass architecture to *Big Brother*: Scenes from a cultural history of transparency. *Cultural Studies Review*, 9(1), pp. 103–23.

Maiorana-Basas, M. and Pagliaro, C. M. (2014). Technology use among adults who are deaf and hard of hearing: A national survey. *Journal of Deaf Studies and Deaf Education*, 19(3), pp. 400–10, https://doi.org/10.1093/deafed/edu005

Manovich, L. (2001). *The Language of New Media*, Cambridge, MA: MIT Press.

Malpas, J. (1999). *Place and Place Experience: A Philosophical Topography*. Cambridge: Cambridge University Press.

Marinetti, F. T. (1991). Tactilism. In *Let's Murder the Moonshine: Selected Writings*. Trans. R. W. Flint and A. A. Coppotelli. Los Angeles: Sun & Moon Classics, pp. 117–20.

Marks, L. U. (2000). *The Skin of Film: Intercultural Cinema, Embodiment and the Senses*. Durham, NC: Duke University Press.

Massey, D. (1994). *Space, Place, and Gender*. Cambridge: Polity Press.

Mauss, M. (1973). Techniques of the body. *Economy and Society*, 2(1), pp. 70–88, https://doi.org/10.1080/03085147300000003

Meese, J. (2014). Google Glass and Australian privacy law: Regulating the future of locative media. In R. Wilken and G. Goggin, eds. *Locative Media*. New York: Routledge, pp. 136–47.

Merleau-Ponty, M. (1964). *The Primacy of Perception, And Other Essays on Phenomenological Psychology*, *the Philosophy of Art, History and Politics*. Edited by J. M. Edie. Trans. W. Cobb. Evanston, IL: Northwestern University Press.

Merleau-Ponty, M. (1968). *The Visible and the Invisible*. Edited by C. Lefort. Trans. A. Lingis. Evanston, IL: Northwestern University Press.

Merleau-Ponty, M. (2004 [1962]). *Phenomenology of Perception*. Trans. C. Smith. London: Routledge.

Mills, M. (2011). On disability and cybernetics: Helen Keller, Norbert Wiener, and the hearing glove. *Differences*, 22, pp. 74–111.

Mills, M. (2014). Cochlear implants after fifty years: A history and an interview with Charles Graser. In S. Gopinath and J. Stanyek, eds. *The Oxford Handbook of Mobile Music Studies, Volume 1*. New York: Oxford University Press, pp. 261–97.

Mitchell, W. J. T. (2005). There are no visual media. *Journal of Visual Culture*, 4(2), pp. 257–66.

Monk, A., Carroll, J., Parker, S. and Blythe, M. (2004). Why are mobile phones annoying? *Behaviour & Information Technology*, 23(1), pp. 33–41.

Moores, S. (2012). *Media, Place and Mobility*. Basingstoke: Palgrave Macmillan.

Moores, S. (2013). We find our way about: Everyday media use and "inhabitant knowledge." *Mobilities*, 10(1), pp. 17–35.

Morley, D. (2003). What's "home" got to do with it? Contradictory dynamics in the domestication of the technology and the dislocation of domesticity. *European Journal of Cultural Studies*, 6(4), pp. 435–58.

Morris, D. (2004). *The Sense of Space*. Albany, NY: State University of New York Press.

Morse, M. (1998). An ontology of everyday distraction: The freeway, the mall, and television. In P. Mellencamp, ed. *Logics of Television: Essays in Cultural Criticism*.

Bloomington: Indiana University Press/London: BFI Publishing, pp. 193–221.

Mukherjee, R. (2020). Sensitivity to electromagnetic stimuli: Entwined histories of wireless signals and plant ecologies. *Media+Environment*, 2(1), https://doi.org/10.1525/001c.13523

Nansen, B. and Wilken, R. (2018). Techniques of the tactile body: A cultural phenomenology of toddlers and mobile touchscreens. *Convergence: The International Journal of New Media Technologies*, 25(1), pp. 60–76.

Neff, G. and Nafus, D. (2016). *Self-Tracking*. Cambridge, MA: MIT Press.

Newton, C. (2016). Snapchat unveils $130 connected sunglasses. *The Verge*, September 24, https://www.theverge.com/2016/9/23/13039184/snapchat-spectacles-price-release-date-snap-inc

Newton, C. (2019). Snap CEO Evan Spiegel on why Spectacles are a new kind of camera. *The Verge*, November 13, https://www.theverge.com/interface/2019/11/13/20960448/evan-spiegel-spectacles-3-interview-snap-snapchat

NIDCD (2015). How do we hear? *National Institute on Deafness and Other Communication Disorders*, March 16, https://nidcd.nih.gov/health/how-do-we-hear

Nist, M. D., Harrison, T. M., Tate, J., Robinson, A., Balas, M. and Pickler, R. H. (2020). Losing touch. *Nursing Inquiry*, 27(3), e12368, https://doi.org/10.1111/nin.12368

Ochoa Gautier, A. M. (2014). *Aurality: Listening and Knowledge in Nineteenth-Century Colombia*. Durham, NC: Duke University Press.

O'Hara, K., Black, A. and Lipson, M. (2006). Everyday practices with mobile video telephony. *CHI 2006 Proceedings, Everyday Use of Mobiles*, April 22–27, 2006, Montreal, Quebec, Canada.

O'Kane, S. (2016). Snap Spectacles review: Fun that's totally worth the trouble. *The Verge*, November 22, https://www.theverge.com/2016/11/21/13671164/snapchat-spectacles-glasses-review-camera-sunglasses

O'Mara, S. (2021). *In Praise of Walking*. New York: W. W. Norton.

O'Neill, M. and Roberts, B. (2020). *Walking Methods: Research on the Move*. New York: Routledge.

Özkan, N. (forthcoming). "Sensing" productivity at home: Self-tracking technologies, gender, and labor in Turkey. *Journal of Computer-Mediated Communication.*

Özkul, D. (2015a). Location as a sense of place: Everyday life, mobile, and spatial practices in urban spaces. In A. de Souza e Silva and M. Sheller, eds. *Mobility and Locative Media: Mobile Communication in Hybrid Spaces*. New York: Routledge, pp. 101–16.

Özkul, D. (2015b). Mobile communication technologies and spatial perception: Mapping London. In R. Wilken and G. Goggin, eds. *Locative Media*. New York: Routledge, pp. 39–51.

Özkul, D. and Humphreys, L. (2015). Record and remember: Memory and meaning-making practices through mobile media. *Mobile Media & Communication*, 3(3), pp. 351–65.

Pallasmaa, J. (2009). *The Thinking Hand: Existential and Embodied Wisdom in Architecture*. Chichester: Wiley.

Parent, L. (2016). The wheeling interview: Mobile methods and disability. *Mobilities*, 11(4), pp. 521–32.

Parikka, J. (2012). New materialism as media theory: medianatures and dirty matter. *Communication and Critical/Cultural Studies*, 9(1), pp. 95–100.

Parisi, D. (2008). Fingerbombing, or "Touching Is Good": The Cultural Construction of Technologized Touch. *The Senses & Society*, 3(3), pp. 302–27.

Parisi, D. (2009). Game interfaces as bodily techniques.

In R. E. Ferdig, ed. *Handbook of Research on Effective Gaming in Education*. New York: IGI Global, pp. 111–26.

Parisi, D. (2015). A counterrevolution in the hands: The console controller as an ergonomic branding mechanism. *Journal of Games Criticism*, 2(1), pp. 1–23, http://gamescriticism.org/articles/parisi-2-1

Parisi, D. (2018). *Archaeologies of Touch: Interfacing with Haptics from Electricity to Computing*. Minneapolis: University of Minnesota Press.

Parisi, D. (2022). Can't touch this. Haptic devices probably won't ever live up to their promise to replicate physical contact. *Real Life*, March 3, https://reallifemag.com/cant-touch-this

Parisi, D., Paterson, M. and Archer, J. E. (2017). Haptic media studies. *New Media & Society*, 19(10), pp. 1513–22.

Passanante Elman, J. (2018). "Find your fit": Wearable technology and the cultural politics of disability. *New Media & Society*, 20(10), pp. 3760–77.

Paterson, M. (2007). *The Senses of Touch: Haptics, Affects and Technologies*. Oxford: Berg.

Paterson, M., Dodge, M. and MacKian, S. (2012). Introduction: Placing touch within social theory and empirical study. In M. Paterson and M. Dodge, eds. *Touching Space, Placing Touch*. Farnham: Ashgate Publishing, pp. 1–28.

Perkins, E. S. (2023). Human eye. Anatomy and physiology. *Britannica*, https://www.britannica.com/science/human-eye

Phone Arena (2013). Breakthrough sensor to give our smartphones a sense of smell. *Phone Arena*, October 8, https://www.phonearena.com/news/Breakthrough-sensor-to-give-our-smartphones-a-sense-of-smell_id48089

PhysOrg (2004). Mobile electronic devices learn to smell. *PhysOrg*, September 16, https://phys.org/news/2004-09-mobile-electronic-devices.html

Pink, S. (2015a). Approaching media through the senses: Between experience and representation. *Media International Australia*, 154, pp. 5–14.

Pink, S. (2015b). *Doing Sensory Ethnography*, 2nd edition. London: Sage.

Plant, S. (2003). On the mobile: The effects of mobile telephones on social and individual life. Archived at https://issaasad.com/wp-content/uploads/2014/08/the-effects-of-mobile-telephones-on-social-and-individual-life.pdf

Powers, D. and Parisi, D. (2020). The hype, haplessness and hope of haptics in the COVID-19 era. *TechCrunch*, July 29, https://techcrunch.com/2020/07/28/the-hype-haplessness-and-hope-of-haptics-in-the-covid-19-era

Puig de la Bellacasa, M. (2017). *Matters of Care: Speculative Ethics in More Than Human Worlds*. Minneapolis: University of Minnesota Press.

Putic, G. (2015). Smartphones about to make leap, carry basic senses. *VOA*, January 28, https://www.voanews.com/a/smartphones-to-carry-basic-sense-of-smell-taste-and-touch/2617420.html

Ramachandran, V. S. and Hubbard, E. M. (2001). Synaesthesia – a window into perception, thought and language. *Journal of Consciousness Studies*, 8(12), pp. 3–34.

Reeves, B. and Nass, C. (1996). *The Media Equation: How People Treat Computers, Television and New Media Like Real People and Places*. Cambridge: Cambridge University Press.

Richardson, I. (2005). Mobile technosoma: Some phenomenological reflections on itinerant media devices. *Fibreculture*, 6, http://six.fibreculturejournal.org

/fcj-032-mobile-technosoma-some-phenomenological-reflections-on-itinerant-media-devices

Richardson, I. (2010). Faces, interfaces, screens: Relational ontologies of attention and distraction. *Transformations*, 18, http://www.transformationsjournal.org/journal/issue_18/article_05.shtml

Richardson, I. (2012). Touching the screen: A phenomenology of mobile gaming and the iPhone. In L. Hjorth, J. Burgess and I. Richardson, eds. *Studying Mobile Media: Cultural Technologies, Mobile Communication, and the iPhone*. New York: Routledge, pp. 133–51.

Richardson, I. and Hjorth, L. (2017). Mobile media, domestic play and haptic ethnography. *New Media & Society*, 19(10), pp. 1653–67.

Richardson, I. and Wilken, R. (2009). Haptic vision, footwork, place-making: A peripatetic phenomenology of the mobile phone pedestrian. *Second Nature: International Journal of Creative Media*, 1(2), https://researchrepository.murdoch.edu.au/id/eprint/11783

Richardson, I. and Wilken, R. (2012). Parerga of the third screen: Mobile media, place and presence. In R. Wilken and G. Goggin, eds. *Mobile Technology and Place*. New York: Routledge, pp. 181–97.

Richardson, I. and Wilken, R. (2017). Mobile media and mediation: The relational ontology of Google Glass. In T. Markham and S. Rodgers, eds. *Conditions of Mediation: Phenomenological Approaches to Media, Technology and Communication*. New York: Peter Lang, pp. 113–23.

Richardson, I., Hjorth, L., Strengers, Y. and Balmford, W. (2017). Careful surveillance at play: Human-animal relations and mobile media in the home. In E. Gómez Cruz, S. Sumartojo and S. Pink, eds. *Refiguring Techniques in Digital Visual Research*. Cham, Switzerland: Palgrave Macmillan, pp. 105–16.

Robson, K. (2022). Google Glass is back after failing hard the first time around – is the world ready? *Verdict*, July 20, https://www.verdict.co.uk/google-testing-augmented-reality-glasses-after-failure-of-google-glass

Romanyshyn, R. D. (1989). *Technology as Symptom and Dream*. London: Routledge.

Ronell, A. (1989). *The Telephone Book: Technology, Schizophrenia, Electric Speech*. Lincoln: University of Nebraska Press.

Rosenthal, R. (ed.) (1969). *McLuhan: Pro and Con*. Baltimore, MD: Penguin.

Rubin, M. L. (1986). Spectacles: Past, present, and future. *Survey of Ophthalmology*, 30(5), pp. 321–27.

Ruthrof, H. (1997). *Semantics and the Body: Meaning from Frege to the Postmodern*. Melbourne: Melbourne University Press.

Rush, J. (2011). Embodied metaphors: Exposing informatic control through first-person shooters. *Games and Culture*, 6(3), pp. 245–58.

Sacks, O. (1985). *The Man Who Mistook His Wife for a Hat and Other Clinical Tales*. New York: Summit Books.

Sacks, O. (2010). *The Mind's Eye*. New York: Alfred A. Knopf Inc.

Saletan, W. (2009). The mind-BlackBerry problem: Hey, you! Cell-phone zombie! Get off the road! *Slate*, October 23, http://www.slate.com/id/2202978

Sanders, R. (2014). Human faces are so variable because we evolved to look unique. *Berkeley News*, September 16, https://news.berkeley.edu/2014/09/16/human-faces-are-so-variable-because-we-evolved-to-look-unique

Sarria, L. C., Han, J., Myrick, J. G. and Potter, R. F. (2022). The effects of "media tech neck": The impact of spinal flexion on cognitive and emotional processing of videos. *ResearchGate* [unpublished paper], https://www.researchgate.net/publication/366464986

_The_Effects_of_'Media_Tech_Neck'_The_Impact_of_Spinal_Flexion_on_Cognitive_and_Emotional_Processing_of_Videos

Sauer, V. J., Eimler, S. C., Maafi, S., Pietrek, M. and Krämer, N. C. (2015). The phantom in my pocket: Determinants of phantom phone sensations. *Mobile Media & Communication*, 3(3), pp. 293–316, https://doi.org/10.1177/2050157914562656

Schafer, R. M. (1977). *The Tuning of the World*. New York: Knopf.

Schloss, J. and Boyer, B. B. (2014). Urban echoes: The boombox and sonic mobility in the 1980s. In S. Gopinath and J. Stanyek, eds. *The Oxford Handbook of Mobile Music Studies, Volume 1*. New York: Oxford University Press, pp. 399–412.

Schwartz, R. (2015). Online place attachment: Exploring technological ties to physical places. In A. de Souza e Silva and M. Sheller, eds. *Mobility and Locative Media: Mobile Communication in Hybrid Spaces*. New York: Routledge, pp. 85–100.

Scientific American. (2015). Ears: Do their design, size and shape matter? *Scientific American*, November 19, https://scientificamerican.com/article/ears-do-their-design-size-and-shape-matter

Sheets-Johnstone, M. (2009). *The Corporeal Turn: An Interdisciplinary Reader*. Exeter: Imprint Academic.

Sherrington, C. S. (1920). *The Integrative Action of the Nervous System*, 6th printing. New Haven, CT: Yale University Press.

Shew, A. (2017). *Animal Constructions and Technological Knowledge*. Lanham, MD: Lexington Books.

Shinkle, E. (2003). Gardens, games and the anamorphic subject: Tracing the body in the virtual landscape. *Digital Arts Conference (DAC): Streaming Worlds*, May 19–23, 2003, Melbourne, Australia.

Silverstone, R. and Hirsch, E., eds. (1992). *Consuming Technologies: Media and Information in Domestic Spaces*. London: Routledge.

Simner, J. (2012). Defining synaesthesia. *British Journal of Psychology*, 103, pp. 1–15.

Simun, M. (2009). My music, my world: Using the mp3 player to shape experience in London. *New Media & Society*, 11(6), pp. 921–41.

Singer, L. (1990). Eye/mind/screen: Toward a phenomenology of cinematic vision. *Quarterly Review of Film and Video*, 12(3), pp. 51–67.

Sitvast, J. (2021). The felt sense and how it can therapeutically be mediated by photographs. *Academia Letters*, Article 665, https://doi.org/10.20935/AL665

Skalski, P., Tamborini, R., Shelton, A., Buncher, M. and Lindmark, P. (2011). Mapping the road to fun: Natural video game controllers, presence, and game enjoyment. *New Media & Society*, 13(2), pp. 224–42.

Sobchack, V. (1992). *The Address of the Eye: A Phenomenology of Film Experience*. Princeton, NJ: Princeton University Press.

Solnit, R. (2000). *Wanderlust: A History of Walking*. New York: Penguin.

Sotamaa, O. (2002). All the world's a Botfighter stage: Notes on location-based multi-user gaming. In F. Mäyrä, ed. *Proceedings of Computer Games and Digital Cultures Conference*. Tampere: Tampere University Press, pp. 35–44.

Srivastava, L. (2005). Mobile phones and the evolution of social behaviour. *Behaviour & Information Technology*, 24(2), pp. 111–29.

Steiner, H. and Veel, K. (2021). *Touch in the Time of Corona: Reflections on Love, Care, and Vulnerability in the Pandemic*. Berlin: De Gruyter.

Stenning, A. (2019). Autism and cognitive embodiment:

Steps towards a non-ableist walking literature. In D. Borthwick, P. Marland and A. Stenning, eds. *Walking, Landscape and Environment*. London: Routledge, pp. 147–67.
Sterne, J. (2006). The mp3 as cultural artifact. *New Media & Society*, 8(5), pp. 825–42.
Stewart, (2007). *Ordinary Affects*. Durham, NC: Duke University Press.
Stromberg, J. (2015). 9 surprising facts about the sense of touch. *Vox*, January 28, https://www.vox.com/2015/1/28/7925737/touch-facts
Subramanian, S. (2021). *How to Feel: The Science and Meaning of Touch*. New York: Columbia University Press.
Sumner, T. D. (2022). Zoom face: Self-surveillance, performance and display. *Journal of Intercultural Studies*, 43(6), pp. 865–79.
Tambornino, J. (2002). *The Corporeal Turn: Passion, Necessity, Politics*. New York: Rowman & Littlefield.
Tiessen, M. (2007). Urban meanderthals and the city of "desire lines." *CTheory.net*, https://journals.uvic.ca/index.php/ctheory/article/view/14515/5358
Turkle, S. (2012). *Alone Together: Why We Expect More From Technology and Less from Each Other*. New York: Basic Books.
Turner, P., Turner, S. and McGregor, I. (2007). Listening, corporeality and presence. In PRESENCE 2007: The 10th Annual International Workshop on Presence, pp. 43–9, http://researchrepository.napier.ac.uk/id/eprint/3488
Twilley, N. (2016). Will smell ever come to smartphones? *The New Yorker*, April 27, https://www.newyorker.com/tech/annals-of-technology/is-digital-smell-doomed
van den Boomen, M. (2014). *Transcoding the Digital: How Metaphors Matter in New Media*. Amsterdam: Institute of Network Cultures.

van der Vorst, E. (2014). A guide to honking in Vietnam. *Saigoneer*, https://saigoneer.com/saigon-culture/2172-a-guide-to-honking-in-vietnam

Vasseleu, C. (1998). *Textures of Light: Vision and Touch in Irigaray, Levinas and Merleau-Ponty*. London: Routledge.

Waldby, C. (1998). Circuits of desire: Internet erotics and the problem of bodily location, https://fremantlestuff.info/readingroom/VID/Circuits3.html

Wall, T. and Webber, N. (2014). Changing cultural coordinates: The transistor radio and space, time, and identity. In S. Gopinath and J. Stanyek, eds. *The Oxford Handbook of Mobile Music Studies, Volume 1*. New York: Oxford University Press, pp. 118–32.

Walters, S. (2014). *Rhetorical Touch: Disability, Identification, Haptics*. Columbia: University of South Carolina Press.

Wang, J. and Tian, J. (2022). The multiple meanings of the ear metaphor in *Hamlet*. *ANQ: A Quarterly Journal of Short Articles, Notes and Reviews*, https://doi.org/10.1080/0895769X.2022.2114413

Ward, A. (2023). *Sensational: A New Story of Our Senses*. London: Profile Books.

Webber, S., Carter, M., Sherwen, S., Smith, W., Joukhadar, Z. and Vetere, F. (2017). Kinecting with orangutans: Zoo visitors' empathetic responses to animals' use of interactive technology. Paper presented at *International Conference on Human Factors in Computing Systems (CHI 17)*, May 6–11, Denver, CO. New York: Association for Computing Machinery (ACM), http://dx.doi.org/10.1145/3025453.3025729

Wee, W. (2014). A mobile app that emits smell, for real. *TechInAsia*, March 6, https://www.techinasia.com/scentee-mobile-app-that-emits-smell

Weibel, P. (1996). The world as interface: Toward the

construction of context-controlled event-worlds. In T. Druckrey, ed. *Electronic Culture: Technology and Visual Representation*. New York: Aperture, pp. 338–51.

Welch, R. B. and Warren, D. H. (1986). Intersensory interactions. In K. R. Boff, L. Kaufman and J. P. Thomas, eds. *Handbook of Perception and Human Performance, Volume 1: Sensory Processes and Perception*. Wiley: New York, pp. 25-1–25-36.

Whatmore, S. (2006). Materialist returns: Practising cultural geography in and for a more-than-human world. *Cultural Geographies*, 13(4), pp. 600–9.

Whyte, W. H. (1980). *The Social Life of Small Urban Spaces*, https://youtu.be/e_5fuaoxHtU

Whyte, W. H. (1988). *City: Rediscovering the Center*. New York: Doubleday.

Wilken, R. (2014). Mobile media, place, and location. In G. Goggin and L. Hjorth, eds. *The Routledge Companion to Mobile Media*. New York: Routledge, pp. 514–27.

Wilken, R. and Goggin, G. (2012). Mobilizing place: Conceptual currents and controversies. In R. Wilken and G. Goggin, eds. *Mobile Technology and Place*. New York: Routledge, pp. 3–25.

Wilken, R. and Humphreys, L. (2019). Constructing the check-in: Reflections on photo-taking among Foursquare users. *Communication and the Public*, 4(2), pp. 100–17.

Wilken, R. and Humphreys, L. (2021). Placemaking through mobile social media platform Snapchat. *Convergence: The International Journal of Research into New Media Technologies*, 27(3), pp. 579–93.

Williamson, J. (1986). *Consuming Passions: The Dynamics of Popular Culture*. New York: Marion Boyars.

Wilmott, C. (2020). *Mobile Mapping: Space, Cartography and the Digital*. Amsterdam: Amsterdam University Press.

Wilmott, C., Fraser, E. and Lammes, S. (2018). "I am he. I am he. Siri rules." Work and play with the Apple Watch. *European Journal of Cultural Studies*, 21(1), pp. 78–95.
Wilson, F. R. (1998). *The Hand: How its Use Shapes the Brain, Language and Human Culture*. New York: Vintage Books.
Worthington, D., Fitch-Hauser, M., Välikoski, T.-R., Imhof, M. and Kim, S.-H. (2011). Listening and privacy management in mobile phone conversations: A cross-cultural comparison of Finnish, German, Korean and United States students. *Empedocles: European Journal for the Philosophy of Communication*, 3(1), pp. 43–60.
Yau, J. M., Olenczak, J. B., Dammann, J. F. and Bensmaia, S. J. (2009). Temporal frequency channels are linked across audition and touch. *Current Biology*, 19(7), pp. 561–66.
Yoon, C. (2018). Assumptions that led to the failure of Google Glass. *Medium*, August 3, https://medium.com/nyc-design/the-assumptions-that-led-to-failures-of-google-glass-8b40a07cfa1e
Zilio, M. (2020). *Faceworld: The Face in the Twenty-First Century*. Cambridge: Polity Press.

Index